ISEE Upper Level Math Preparation

www.EffortlessMath.com

… So Much More Online!

✓ FREE Math lessons

✓ More Math learning books!

✓ Mathematics Worksheets

✓ Online Math Tutors

Need a PDF version of this book?

Send email to: info@EffortlessMath.com

ISEE Upper Level Math Preparation Exercise Book

A Comprehensive Math Workbook and Two Full-Length ISEE Upper Level Math Practice Tests

By

Reza Nazari

& Sam Mest

Copyright © 2019

Reza Nazari & Sam Mest

All rights reserved. No part of this publication may be reproduced, stored in a retrieval system, or transmitted in any form or by any means, electronic, mechanical, photocopying, recording, scanning, or otherwise, except as permitted under Section 107 or 108 of the 1976 United States Copyright Ac, without permission of the author.

All inquiries should be addressed to:

info@effortlessMath.com

www.EffortlessMath.com

ISBN-13: 978-1-64612-000-0

ISBN-10: 1-64612-000-0

Published by: Effortless Math Education

www.EffortlessMath.com

Description

ISEE Upper Level Math Preparation Exercise Book provides test takers with an in-depth focus on the math portion of the exam, helping them master the math skills that students find the most troublesome. It is designed to address the needs of ISEE Upper Level test takers who must have a working knowledge of basic Math.

This comprehensive ISEE Upper Level Math Workbook contains many exciting and unique features to help you score higher on the ISEE Upper Level Math test, including:

- Content 100% aligned with the 2019 ISEE Upper Level test
- Prepared by ISEE Upper Level Math experts
- Complete coverage of all ISEE Upper Level Math topics which you will need to ace the test
- Over 2,500 additional ISEE Upper Level math practice questions with answers
- 2 complete ISEE Upper Level Math practice tests (featuring new question types) with detailed answers

ISEE Upper Level Math Preparation Exercise Book is an incredibly useful tool for those ISEE Upper Level test takers who want to review core content areas, brush-up in math, discover their strengths and weaknesses, and achieve their best scores on the ISEE Upper Level test.

Contents

Section 1: Arithmetic 8
   Chapter 1: Fractions and Decimals 9
      Simplifying Fractions 10
      Adding and Subtracting Fraction 11
      Multiplying and Dividing Fractions 12
      Adding Mixed Numbers 13
      Subtract Mixed Numbers 14
      Multiplying Mixed Numbers 15
      Dividing Mixed Numbers 16
      Comparing Decimals 17
      Rounding Decimals 18
      Adding and Subtracting Decimals 19
      Multiplying and Dividing Decimals 20
      Converting Between Fractions, Decimals and Mixed Numbers 21
      Factoring Numbers 22
      Greatest Common Factor 23
      Least Common Multiple 24
      Answers of Worksheets 25
   Chapter 2: Real Numbers and Integers 30
      Adding and Subtracting Integers 31
      Multiplying and Dividing Integers 32
      Ordering Integers and Numbers 33
      Arrange, Order, and Comparing Integers 34
      Order of Operations 35
      Mixed Integer Computations 36
      Integers and Absolute Value 37
      Answers of Worksheets 38
   Chapter 3: Proportions and Ratios 41
      Writing Ratios 42
      Simplifying Ratios 43
      Create a Proportion 44

Similar Figures 45

Simple Interest 46

Ratio and Rates Word Problems 47

Answers of Worksheets 48

Chapter 4: Percent 50

Percentage Calculations 51

Converting Between Percent, Fractions, and Decimals 52

Percent Problems 53

Find What Percentage a Number Is of Another 54

Find a Percentage of a Given Number 55

Percent of Increase and Decrease 56

Markup, Discount, and Tax 57

Answers of Worksheets 58

Section 2: Algebra 60

Chapter 5: Algebraic Expressions 61

Expressions and Variables 62

Simplifying Variable Expressions 63

Simplifying Polynomial Expressions 64

The Distributive Property 65

Evaluating One Variable 66

Evaluating Two Variables 67

Combining like Terms 68

Answers of Worksheets 69

Chapter 6: Equations and Inequalities 72

One–Step Equations 73

One–Step Equation Word Problems 74

Two–Step Equations 75

Two–Step Equation Word Problems 76

Multi–Step Equations 77

Graphing Single–Variable Inequalities 78

One–Step Inequalities 79

Multi-Step Inequalities 80

Answers of Worksheets 81

Chapter 7: Systems of Equations 84
    Solving Systems of Equations by Substitution 85
    Solving Systems of Equations by Elimination 86
    Systems of Equations Word Problems 87
    Answers of Worksheets 88

Chapter 8: Linear Functions 89
    Finding Slope 90
    Graphing Lines Using Slope–Intercept Form 91
    Graphing Lines Using Standard Form 92
    Writing Linear Equations 93
    Graphing Linear Inequalities 94
    Finding Midpoint 95
    Finding Distance of Two Points 96
    Slope and Rate of Change 97
    Find the Slope, x–intercept and y–intercept 98
    Write an equation from a graph 99
    Slope–intercept Form 100
    Point–slope Form 101
    Equations of Horizontal and Vertical Lines 102
    Equation of Parallel or Perpendicular Lines 103
    Answers of Worksheets 104

Chapter 9: Monomials and Polynomials 111
    Writing Polynomials in Standard Form 112
    Simplifying Polynomials 113
    Add and Subtract monomials 114
    Multiplying Monomials 115
    Multiplying and Dividing Monomials 116
    GCF of Monomials 117
    Powers of monomials 118
    Multiplying a Polynomial and a Monomial 119
    Multiplying Binomials 120
    Factoring Trinomials 121
    Answers of Worksheets 122

Chapter 10: Exponents and Radicals ... 128
Multiplication Property of Exponents ... 129
Division Property of Exponents ... 130
Powers of Products and Quotients ... 131
Zero and Negative Exponents ... 132
Writing Scientific Notation ... 133
Square Roots ... 134
Answers of Worksheets ... 135

Section 3: Geometry and Statistics ... 138

Chapter 11: Plane Figures ... 139
Transformations: Translations, Rotations, and Reflections ... 140
The Pythagorean Theorem ... 141
Area of Triangles ... 142
Perimeter of Polygon ... 143
Area and Circumference of Circles ... 144
Area of Squares, Rectangles, and Parallelograms ... 145
Area of Trapezoids ... 146
Answers of Worksheets ... 147

Chapter 12: Solid Figures ... 149
Volume of Cubes and Rectangle Prisms ... 150
Surface Area of Cubes ... 151
Surface Area of a Prism ... 152
Volume of Pyramids and Cones ... 153
Answers of Worksheets ... 154

Chapter 13: Statistics ... 155
Mean, Median, Mode, and Range of the Given Data ... 156
Box and Whisker Plot ... 157
Bar Graph ... 158
Stem–And–Leaf Plot ... 159
The Pie Graph or Circle Graph ... 160
Scatter Plots ... 161
Answers of Worksheets ... 162

Chapter 14: Probability ... 168

- Probability of Simple Events 169
- Experimental Probability.. 170
- Factorials 171
- Combination and Permutations 172
- Answers of Worksheets 173

Time to Test 174
- ISEE Upper Level Mathematics Practice Test 1 178
- ISEE UPPER LEVEL Mathematics Practice Test 2 189

ISEE Upper Level Practice Tests Answers and Explanations 198
- ISEE Upper Level Practice Test 2 Answers 199
- ISEE Upper Level Practice Test 2 Answers 200
- ISEE Upper Level Practice Tests 1 Answers and Explanations 201
- ISEE Upper Level Practice Tests 2 Answers and Explanations 206

Section 1: Arithmetic

- *Fractions and Decimals*
- *Real Numbers and Integers*
- *Proportions and Ratios*
- *Percent*

Chapter 1:

Fractions and Decimals

Topics that you'll learn in this part:

- ✓ Simplifying Fractions
- ✓ Adding and Subtracting Fractions
- ✓ Multiplying and Dividing Fractions
- ✓ Adding Mixed Numbers
- ✓ Subtract Mixed Numbers
- ✓ Multiplying Mixed Numbers
- ✓ Dividing Mixed Numbers
- ✓ Comparing Decimals
- ✓ Rounding Decimals
- ✓ Adding and Subtracting Decimals
- ✓ Multiplying and Dividing Decimals
- ✓ Converting Between Fractions, Decimals and Mixed Numbers
- ✓ Factoring Numbers
- ✓ Greatest Common Factor
- ✓ Least Common Multiple
- ✓ Divisibility Rules

Simplifying Fractions

Simplify the fractions.

1) $\frac{33}{54} =$ 11/18

2) $\frac{12}{15} =$

3) $\frac{18}{27} =$

4) $\frac{12}{16} =$

5) $\frac{26}{78} =$

6) $\frac{10}{40} =$

7) $\frac{20}{45} =$

8) $\frac{18}{36} = \frac{1}{2}$

9) $\frac{40}{100} = \frac{2}{5}$

10) $\frac{6}{54} = \frac{1}{9}$

11) $\frac{15}{27} = \frac{5}{9}$

12) $\frac{15}{20} = \frac{3}{4}$

13) $\frac{20}{32} = \frac{5}{8}$

14) $\frac{26}{32} = \frac{13}{16}$

15) $\frac{15}{75} = \frac{1}{5}$

16) $\frac{40}{70} = \frac{4}{7}$

17) $\frac{24}{48} = \frac{1}{2}$

18) $\frac{35}{84} = \frac{5}{12}$

19) $\frac{15}{40} = \frac{3}{8}$

20) $\frac{15}{60} = \frac{3}{12} = \frac{1}{4}$

21) $\frac{30}{54} = \frac{10}{18} = \frac{5}{9}$

Solve

22) Which of the following fractions equal to $\frac{4}{5}$? _____

    A. $\frac{64}{75}$ B. $\frac{92}{115}$ C. $\frac{60}{85}$ D. $\frac{160}{220}$

23) Which of the following fractions equal to $\frac{3}{7}$? _____

    A. $\frac{63}{147}$ B. $\frac{75}{182}$ C. $\frac{54}{140}$ D. $\frac{39}{98}$

24) Which of the following fractions equal to $\frac{7}{15}$? _____

    A. $\frac{33}{56}$ B. $\frac{25}{85}$ C. $\frac{42}{90}$ D. $\frac{23}{72}$

Adding and Subtracting Fraction

✎ Add fractions.

1) $\frac{3}{5} + \frac{2}{4} =$

2) $\frac{2}{7} + \frac{3}{5} =$

3) $\frac{2}{7} + \frac{1}{4} =$

4) $\frac{7}{8} + \frac{5}{3} =$

5) $\frac{3}{5} + \frac{1}{10} =$

6) $\frac{2}{9} + \frac{2}{3} =$

7) $\frac{2}{4} + \frac{2}{7} =$

8) $\frac{4}{3} + \frac{1}{4} =$

9) $\frac{9}{21} + \frac{3}{7} =$

✎ Subtract fractions.

10) $\frac{3}{5} - \frac{1}{5} =$

11) $\frac{4}{5} - \frac{3}{6} =$

12) $\frac{1}{3} - \frac{1}{6} =$

13) $\frac{5}{8} - \frac{1}{5} =$

14) $\frac{4}{5} - \frac{6}{10} =$

15) $\frac{12}{20} - \frac{3}{10} =$

16) $\frac{5}{6} - \frac{8}{18} =$

17) $\frac{5}{12} - \frac{18}{24} =$

18) $\frac{1}{5} - \frac{1}{8} =$

✎ Solve.

19) A city worker is painting a stripe down the center of Main Street. Main Street is $\frac{8}{10}$ mile long. The worker painted $\frac{3}{10}$ mile of the street. How much of the street painting is left?

20) From a board 8 feet in length, Tim cut to $2\frac{1}{3}$ foot book shelves. How much of the board remained?

21) While taking inventory at his pastry shop, Eddie realizes that he had $\frac{1}{2}$ of a box of baking powder yesterday, but the supply is now down to $\frac{1}{8}$ of a box. How much more baking powder did Eddie have yesterday?

Multiplying and Dividing Fractions

✎ Multiply fractions. Then simplify.

1) $\frac{2}{7} \times \frac{5}{9} =$

2) $\frac{2}{4} \times \frac{3}{7} =$

3) $\frac{1}{2} \times \frac{4}{7} =$

4) $\frac{2}{3} \times \frac{1}{5} =$

5) $\frac{9}{10} \times \frac{2}{3} =$

6) $\frac{5}{7} \times \frac{9}{11} =$

7) $\frac{6}{9} \times \frac{2}{6} =$

8) $\frac{3}{4} \times \frac{2}{5} =$

9) $\frac{4}{7} \times \frac{7}{9} =$

✎ Divide fractions. Simplify if necessary.

10) $\frac{1}{9} \div \frac{3}{5} =$

11) $\frac{1}{6} \div \frac{2}{7} =$

12) $\frac{6}{9} \div \frac{7}{9} =$

13) $\frac{12}{9} \div \frac{13}{8} =$

14) $\frac{3}{20} \div \frac{5}{10} =$

15) $\frac{1}{5} \div \frac{3}{2} =$

16) $\frac{4}{6} \div \frac{3}{6} =$

17) $\frac{11}{24} \div \frac{1}{12} =$

18) $\frac{7}{14} \div \frac{3}{9} =$

19) $\frac{35}{14} \div \frac{7}{14} =$

20) $\frac{63}{15} \div \frac{14}{20} =$

21) $\frac{110}{50} \div \frac{11}{20} =$

✎ Solve.

22) Vera is using her phone. Its battery life is down to $\frac{2}{5}$, and it drains another $\frac{1}{9}$ every hour. How many hours will her battery last?

   A. $\frac{25}{9}$ B. $\frac{18}{5}$ C. $\frac{16}{5}$ D. 5

23) A factory uses $\frac{1}{3}$ of a barrel of raisins in each batch of granola bars. Yesterday, the factory used $\frac{2}{3}$ of a barrel of raisins. How many batches of granola bars did the factory make yesterday?

   A. $\frac{1}{3}$ B. $\frac{2}{3}$ C. $\frac{3}{2}$ D. 2

Adding Mixed Numbers

✏️ **Add.**

1) $2\frac{1}{4} + 1\frac{1}{2} =$

2) $3\frac{1}{5} + 2\frac{3}{5} =$

3) $2\frac{2}{7} + 1\frac{1}{7} =$

4) $2\frac{1}{4} + 1\frac{3}{4} =$

5) $2\frac{1}{5} + 4\frac{1}{10} =$

6) $3\frac{2}{5} + 1\frac{3}{5} =$

7) $2\frac{1}{5} + 2\frac{2}{3} =$

8) $3\frac{1}{6} + 5\frac{1}{2} =$

9) $3\frac{3}{7} + 5\frac{4}{7} =$

10) $3 + \frac{1}{3} =$

11) $2\frac{2}{5} + \frac{1}{3} =$

12) $2\frac{1}{3} + 2\frac{1}{9} =$

✏️ **Solve.**

13) A baker used $4\frac{1}{2}$ bags of floor baking cakes and $3\frac{3}{5}$ bags of floor baking cookies. How much floor did he used in all?

A. $10\frac{1}{10}$ B. $8\frac{1}{10}$ C. $\frac{16}{10}$ D. $6\frac{2}{10}$

14) Sam bought $2\frac{1}{2}$ kg of sugar from one shop and $6\frac{2}{3}$ kg of sugar from the other shop. How much sugar did he buy in all?

A. $9\frac{1}{6}$ B. $8\frac{1}{6}$ C. $5\frac{1}{6}$ D. $6\frac{1}{6}$

15) A tank has $82\frac{3}{4}$ liters of water. $24\frac{4}{5}$ liters were used and the tank was filled with another $18\frac{3}{4}$ liters. What is the final volume of water in the tank?

A. $75\frac{1}{10}$ B. $70\frac{7}{10}$ C. $76\frac{7}{10}$ D. $76\frac{1}{10}$

Subtract Mixed Numbers

✏ Subtract.

1) $3\frac{2}{3} - 2\frac{1}{3} =$

2) $8\frac{1}{2} - 3\frac{1}{6} =$

3) $6\frac{1}{5} - 3\frac{4}{5} =$

4) $3\frac{1}{7} - 2\frac{2}{7} =$

5) $4\frac{1}{4} - 2\frac{1}{3} =$

6) $5\frac{1}{2} - 3\frac{1}{6} =$

7) $6\frac{1}{4} - 3\frac{1}{2} =$

8) $2\frac{2}{4} - 2\frac{1}{2} =$

9) $2\frac{7}{8} - 3\frac{2}{5} =$

10) $3\frac{1}{3} - 2\frac{2}{6} =$

11) $8\frac{1}{23} - 3\frac{1}{23} =$

12) $3\frac{1}{4} - \frac{7}{12} =$

✏ Solve.

13) When Frodo smiles, his mouth is $2\frac{3}{4}$ in wide. When he is not smiling, his mouth is only $2\frac{1}{4}$ in wide. How much wider is Frodo's mouth when he is smiling than when he is not smiling?

14) Jack jumped $4\frac{1}{7}$ m in a long jump competition. Shane jumped $3\frac{2}{9}$ m. Who jumped longer and by how many meters?

15) Sharon spent $4\frac{3}{7}$ hours studying math and playing tennis. If she played tennis for $2\frac{1}{2}$ hours, how long did she study?

16) Ed just filled up at the gas station, and now his car fuel gauge reads $\frac{8}{10}$ full. He didn't fill the gas tank. If the gauge of fuel was at $\frac{4}{10}$ when he got to the gas station, what fraction of the tank did he fill at the gas station?

Multiplying Mixed Numbers

✍ *Find each product.*

1) $2\frac{3}{4} \times 1\frac{1}{3} =$

2) $2\frac{2}{7} \times 2\frac{4}{5} =$

3) $9\frac{1}{2} \times 3\frac{1}{3} =$

4) $3\frac{1}{5} \times 3\frac{1}{3} =$

5) $4\frac{3}{7} \times 2\frac{5}{8} =$

6) $1\frac{1}{7} \times 2\frac{5}{7} =$

7) $2\frac{1}{2} \times 3\frac{1}{4} =$

8) $1\frac{1}{9} \times 3\frac{1}{2} =$

9) $4\frac{2}{3} \times \frac{3}{7} =$

10) $5\frac{3}{5} \times 2\frac{1}{2} =$

11) $3\frac{1}{5} \times 3\frac{1}{2} =$

12) $4\frac{1}{3} \times 1\frac{1}{3} =$

✍ *Solve.*

13) Kerry read $\frac{2}{3}$ of her chemistry book containing 420 pages. David read $\frac{3}{4}$ of the same. Who read more pages and by how many pages?

14) Victor's weight was 60 kg. He lost $\frac{1}{10}$ of his weight in 3 months. How much weight did he lost?

15) Alex bought 15 kg sweets on his birthday and distributed $\frac{3}{4}$ of it among his friends. How much sweets did he distribute?

16) Shelly distributed a fraction of a cake among 6 girls. Each girl got $\frac{1}{9}$ part of the cake. What fraction of the cake did she distribute in all?

17) The elephants at the Pike Zoo are fed $\frac{1}{2}$ of a barrel of corn each day. The buffalo are fed $\frac{9}{10}$ as much corn as the elephants. How many barrels of corn are the buffalo fed each day?

Dividing Mixed Numbers

✎ Find each quotient.

1) $3\frac{1}{5} \div 2\frac{2}{3} =$

2) $4\frac{2}{7} \div 3\frac{1}{2} =$

3) $3\frac{1}{3} \div 4\frac{2}{7} =$

4) $3\frac{3}{7} \div 7\frac{1}{3} =$

5) $3\frac{3}{4} \div 1\frac{3}{5} =$

6) $2\frac{7}{8} \div 2\frac{2}{7} =$

7) $1\frac{1}{2} \div 1\frac{2}{5} =$

8) $3\frac{1}{3} \div 2\frac{1}{3} =$

9) $7\frac{1}{7} \div 3\frac{4}{7} =$

10) $3\frac{4}{5} \div 6\frac{1}{3} =$

11) $4\frac{2}{7} \div 7\frac{2}{4} =$

12) $3\frac{1}{5} \div 1\frac{2}{10} =$

13) $1\frac{2}{3} \div 3\frac{1}{3} =$

14) $2\frac{1}{4} \div 1\frac{1}{2} =$

15) $10\frac{1}{2} \div 1\frac{2}{3} =$

16) $3\frac{1}{6} \div 4\frac{2}{3} =$

17) $4\frac{1}{8} \div 2\frac{1}{2} =$

18) $2\frac{1}{10} \div 2\frac{3}{5} =$

19) $1\frac{4}{11} \div 1\frac{1}{4} =$

20) $9\frac{1}{2} \div 9\frac{2}{3} =$

21) $8\frac{3}{4} \div 2\frac{2}{5} =$

22) $12\frac{1}{2} \div 9\frac{1}{3} =$

23) $2\frac{1}{8} \div 1\frac{1}{2} =$

24) $1\frac{1}{10} \div 1\frac{3}{5} =$

✎ Solve.

25) The product of two numbers is 18. If one number is $8\frac{2}{5}$ find the other number.

26) Ashish cut a 25 m long rope into pieces of $1\frac{2}{3}$ meters each. Find the total number of pieces he cut.

27) The cost of $5\frac{2}{5}$ kg of sugar is $\$101\frac{1}{4}$ find its cost per kg.

28) Ana drinks chocolate milk out of glasses that each hold $\frac{1}{8}$ of a liter. She has $\frac{7}{10}$ of a liter of chocolate milk in her refrigerator. How many glasses of chocolate milk can she pour?

Comparing Decimals

✎ Write the correct comparison symbol (>, < or =).

1) 0.632 ☐ 0.631
2) 0.75 ☐ 1
3) 3.91 ☐ 4.91
4) 3.2 ☐ 3.1
5) 2.8 ☐ 2.801
6) 0.4 ☐ 0.74
7) 14.9 ☐ 1.49
8) 0.707 ☐ 0.0707

9) 1.0001 ☐ 0.999
10) 3.655 ☐ 6.6555
11) 15.4 ☐ 14.5
12) 0.909 ☐ 0.99
13) 3.3 ☐ 3.33
14) 0.304 ☐ 0.304
15) 4.0001 ☐ 4.001
16) 3.003 ☐ 3.3

17) 2.85 ☐ 2.88
18) 0.98 ☐ 0.908
19) 2.031 ☐ 2.0031
20) 5.97 ☐ 5.79
21) 6.302 ☐ 6.203
22) 0.075 ☐ 0.57
23) 1.04 ☐ 1.0401
24) 9.101 ☐ 9.011

✎ Solve.

25) Write the following decimals in ascending order:

$$5.64, 2.54, 3.05, 0.259 \text{ and } 8.32$$

26) Abril wants a cold drink to take with her to the park. She is choosing between a bottle of sparkling water that contains 502.75 ml and a bottle of plain water that contains 499.793 ml. It is a hot day and Abril wants to bring as much to drink as possible. Which beverage should Abril choose?

Rounding Decimals

Round each decimal number to the nearest place indicated.

1) 3.<u>1</u>2
2) 0.3<u>5</u>6
3) 0.<u>4</u>9
4) <u>6</u>.75
5) 1.7<u>2</u>4
6) 3.2<u>7</u>6
7) 3.<u>3</u>45
8) <u>10</u>.66
9) <u>44</u>.93
10) 7.<u>0</u>51
11) 12.6<u>4</u>6
12) <u>7</u>.46
13) 5.<u>8</u>63
14) <u>101</u>.03
15) <u>1</u>.53
16) 0.<u>3</u>51
17) <u>100</u>.45
18) <u>7</u>.77

Round off the following to the nearest tenths.

19) 22.652
20) 30.342
21) 47.847
22) 82.88
23) 16.184
24) 71.79

Round off the following to the nearest hundredths.

25) 5.439
26) 12.907
27) 26.1855
28) 48.623
29) 91.448
30) 29.354

Round off the following to the nearest whole number.

31) 23.18
32) 8.6
33) 14.45
34) 7.5
35) 3.95
36) 56.7
37) 13.75
38) 12.55
39) 14.25
40) 156
41) 6.52
42) 12.34
43) 50.51
44) 31.501
45) 101.16

Adding and Subtracting Decimals

✏ Add and subtract decimals.

1) 17.81 − 10.38

2) 37.03 − 15.9

3) 64.12 − 33.33

4) 14.58 + 15.03

5) 17.96 + 10.01

6) 43.02 + 71.08

7) 93.09 − 66.18

8) 76.36 − 52.60

9) 98.45 + 45.56

10) 12.5 + 11.11

11) 34.02 − 39.00

12) 17.56 + 13.98

✏ Solve.

13) $3.56 + \square = 14.7$

14) $\square + 3.5 = 10.6$

15) $12.46 + \square = 17.18$

16) $\square + 6.39 = 10.8$

17) $\square + 3.25 = 5$

18) $14.2 + \square = 17.85$

✏ Solve.

19) Henry weighed two colored metal balls during a science class. The yellow ball weighed 0.88 pounds and the green ball weighed 0.47 pounds. If Henry places both balls on the scale at the same time, what will the scale read?

20) Scarlett has a piece of brown ribbon that is 3.41 inches long and a piece of orange ribbon that is 2.22 inches long. How much longer is the brown ribbon?

21) Kate had $368.29. Her mother gave her $253.46 and her sister gave her $57.39. How much money does she has now?

Multiplying and Dividing Decimals

✏ Find each product.

1) 6.5 × 3.3

2) 4.7 × 7.4

3) 0.99 × 1.85

4) 71.5 × 0.55

5) 33.1 × 3.75

6) 44.3 × 3.31

7) 47.3 × 14.9

8) 15.6 × 14.1

9) 3.75 × 5.41

✏ Solve.

10) A hose in a dessert factory pumps out 9.8 liters of chocolate syrup each minute. How many liters of chocolate syrup will the hose pump out in 8 minutes?

11) Diana has a set of wooden boards. Each board is 5.3 meters long. If Diana lays 8 boards end-to-end, how many meters long will the line of boards be?

✏ Find each quotient.

12) $19.5 \div 6.2 =$

13) $45.1 \div 5.5 =$

14) $35.5 \div 10.5 =$

15) $12.5 \div 3.2 =$

16) $17.7 \div 10.3 =$

17) $19.9 \div 20.1 =$

18) $33.8 \div 9.3 =$

19) $71.1 \div 25.3 =$

20) $50.3 \div 40.1 =$

✏ Solve.

21) A factory used 96.7 kilograms of tomatoes to make 4 batches of pasta sauce. What quantity of tomatoes did the factory put in each batch?

Converting Between Fractions, Decimals and Mixed Numbers

✎ **Convert fractions to decimals.**

1) $\frac{7}{10} =$

2) $\frac{15}{25} =$

3) $\frac{6}{18} =$

4) $\frac{3}{8} =$

5) $\frac{12}{48} =$

6) $\frac{21}{7} =$

7) $\frac{35}{10} =$

8) $\frac{75}{15} =$

9) $\frac{66}{10} =$

✎ **Solve.**

10) Maria and Darcy are in the same math class. Maria has completed $\frac{2}{3}$ of her math homework. Darcy has completed $\frac{5}{6}$ of her math homework. Which girl has completed more of her math homework?

11) For track practice, runners were supposed to walk or jog twenty laps. Sara jogged $\frac{3}{4}$ of the laps. Jacob jogged $\frac{3}{5}$ of the laps. Sierra jogged $\frac{1}{2}$ of the laps. List the runners in order from least to greatest number of laps jogged.

12) At a sports banquet, Garrett ate $\frac{5}{6}$ of a pizza. Massey ate $1\frac{1}{3}$ of a pizza. Jaime ate $\frac{1}{2}$ of a pizza. List the students in order from who ate the least to who ate the most.

✎ **Convert decimal into fraction or mixed numbers.**

13) 0.7

14) 0.25

15) 4.3

16) 9.25

17) 4.7

18) 15.5

Factoring Numbers

🖎 List all positive factors of each number.

1) 35
2) 17
3) 42
4) 24
5) 33
6) 22
7) 39
8) 51
9) 34
10) 18
11) 69
12) 58
13) 76
14) 48
15) 14
16) 12
17) 18
18) 20
19) 24
20) 72
21) 85
22) 38
23) 35
24) 8
25) 5
26) 4
27) 36
28) 42
29) 56
30) 63
31) 80
32) 95
33) 102

🖎 List the prime factorization for each number.

34) 30
35) 56
36) 78
37) 25
38) 46
39) 28
40) 63
41) 52
42) 18
43) 48
44) 58
45) 36
46) 124
47) 90
48) 69
49) 72
50) 85
51) 21
52) 12
53) 18
54) 24
55) 55
56) 75
57) 9
58) 10
59) 15
60) 14

Greatest Common Factor

✎ **Find the GCF for each number pair.**

1) 36, 12
2) 18, 34
3) 25, 55
4) 18, 48
5) 21, 90

6) 14, 42
7) 15, 80
8) 10, 35
9) 70, 30
10) 28, 36

11) 17, 34
12) 54, 14
13) 39, 24
14) 30, 65
15) 72, 20

✎ **Solve.**

16) Sara has 16 red flowers and 24 yellow flowers. She wants to make a bouquet with the same of each color flower in each bouquet. What is the greatest number of bouquets she can make?

17) At a concert, the band has 8 men's T-shirts and 16 women's T-shirts. The band wants to set up tables to sell the shirts, with an equal number of men's and women's shirts available at each table and no shirts left over. What is the greatest number of tables the band can sell shirts from?

18) Nancy is planting 6 bushes and 15 trees in rows. If she wants all the rows to be the same, with no plants left over, what is the greatest number of rows Nancy can plant?

19) Peter has 12 dollars in his pocket and James has 15 dollars. They want to give money to each other. How much money will they have left after they give to each other the same but highest possible amount?

Least Common Multiple

✏ Find the LCM for each number pair.

1) 25,10
2) 36,18
3) 8,10
4) 12,18
5) 24,32
6) 14,10
7) 8,28
8) 51,57
9) 20,15,10
10) 12,20,28
11) 15,75
12) 10,25
13) 9,7
14) 78,6
15) 22,10,2
16) 12,4,16
17) 9,21
18) 25,15,20
19) 70,10
20) 12,18,24
21) 15,45,30

✏ Solve.

22) Becky is packing equal quantities of pretzels and crackers for snacks. Becky bags the pretzels in groups of 4 and the crackers in groups of 18. What is the smallest number of crackers that she can pack?

23) Sam and Carlos are bowling with plastic pins in Sam's living room. Remarkably, Sam knocks down 8 pins on every bowl, and Carlos knocks down 9 pins on every bowl. At the end of the day, Sam and Carlos have knocked down the same total number of pins. What is the least number of total pins that Sam and Carlos could have each knocked down?

24) Regan's Bakery sells muffins in packages of 9 and cookies in packages of 11. Going through yesterday's receipts, a store manager notices that the bakery sold the same number of muffins and cookies yesterday afternoon. What is the smallest number of muffins that the bakery could have sold?

Answers of Worksheets

Simplifying Fractions

1) $\frac{11}{18}$
2) $\frac{4}{5}$
3) $\frac{2}{3}$
4) $\frac{3}{4}$
5) $\frac{1}{3}$
6) $\frac{1}{4}$
7) $\frac{4}{9}$
8) $\frac{1}{2}$
9) $\frac{2}{5}$
10) $\frac{1}{9}$
11) $\frac{5}{9}$
12) $\frac{3}{4}$
13) $\frac{5}{8}$
14) $\frac{13}{16}$
15) $\frac{1}{5}$
16) $\frac{4}{7}$
17) $\frac{1}{2}$
18) $\frac{5}{12}$
19) $\frac{3}{8}$
20) $\frac{1}{4}$
21) $\frac{5}{9}$
22) B
23) A
24) C

Adding and Subtracting Fractions

1) $\frac{11}{10}$
2) $\frac{31}{35}$
3) $\frac{15}{28}$
4) $\frac{61}{24}$
5) $\frac{7}{10}$
6) $\frac{8}{9}$
7) $\frac{11}{14}$
8) $\frac{19}{12}$
9) $\frac{18}{21}$
10) $\frac{2}{5}$
11) $\frac{3}{10}$
12) $\frac{1}{6}$
13) $\frac{17}{40}$
14) $\frac{1}{5}$
15) $\frac{3}{10}$
16) $\frac{7}{18}$
17) $-\frac{1}{3}$
18) $\frac{3}{40}$
19) $\frac{1}{2}$
20) $5\frac{2}{3}$
21) $\frac{3}{8}$

Multiplying and Dividing Fractions

1) $\frac{10}{63}$
2) $\frac{3}{14}$
3) $\frac{2}{7}$
4) $\frac{2}{15}$
5) $\frac{3}{5}$
6) $\frac{45}{77}$
7) $\frac{2}{9}$
8) $\frac{3}{10}$
9) $\frac{4}{9}$
10) $\frac{5}{27}$
11) $\frac{7}{12}$
12) $\frac{6}{7}$
13) $\frac{32}{39}$
14) $\frac{3}{10}$
15) $\frac{2}{15}$
16) $\frac{4}{3}$
17) $\frac{11}{2}$
18) $\frac{3}{2}$
19) 5
20) 6
21) 4
22) $\frac{18}{5}$
23) 2

Adding Mixed Numbers

1) $3\frac{3}{4}$
2) $5\frac{4}{5}$
3) $3\frac{3}{7}$
4) 4
5) $6\frac{3}{10}$
6) 5
7) $4\frac{13}{15}$
8) $8\frac{2}{3}$
9) 9
10) $3\frac{1}{3}$
11) $2\frac{11}{15}$
12) $4\frac{4}{9}$
13) $8\frac{1}{10}$
14) $9\frac{1}{6}$
15) $76\frac{7}{10}$

Subtract Mixed Numbers

1) $1\frac{1}{3}$
2) $5\frac{1}{3}$
3) $2\frac{2}{5}$
4) $\frac{6}{7}$
5) $1\frac{11}{12}$
6) $2\frac{1}{3}$
7) $2\frac{3}{4}$
8) 0
9) $-\frac{21}{40}$
10) 1
11) 5
12) $2\frac{2}{3}$
13) $\frac{1}{2}$
14) $\frac{58}{63}$, Jack
15) $1\frac{13}{14}$
16) $\frac{2}{5}$

Multiplying Mixed Numbers

1) $3\frac{2}{3}$
2) $6\frac{2}{5}$
3) $31\frac{2}{3}$
4) $10\frac{2}{3}$
5) $11\frac{5}{8}$
6) $21\frac{5}{7}$
7) $8\frac{1}{8}$
8) $3\frac{8}{9}$
9) 2
10) 14
11) $11\frac{1}{5}$
12) $5\frac{7}{9}$
13) David, 315 pages
14) 6kg
15) $11\frac{1}{4}$
16) $\frac{6}{9}$
17) $\frac{9}{20}$

Dividing Mixed Numbers

1) $1\frac{1}{5}$
2) $1\frac{11}{49}$
3) $\frac{7}{9}$
4) $\frac{36}{77}$
5) $2\frac{11}{32}$
6) $1\frac{33}{128}$
7) $1\frac{1}{14}$
8) $1\frac{3}{7}$
9) 2
10) $\frac{3}{5}$
11) $\frac{4}{7}$
12) $2\frac{2}{3}$
13) $\frac{1}{2}$
14) $1\frac{1}{2}$
15) $6\frac{3}{10}$
16) $\frac{19}{28}$
17) $1\frac{13}{20}$
18) $\frac{21}{26}$
19) $1\frac{1}{11}$
20) $\frac{57}{58}$
21) $3\frac{31}{48}$
22) $1\frac{19}{56}$
23) $1\frac{5}{12}$
24) $\frac{11}{16}$
25) $2\frac{1}{7}$
26) 15
27) $18\frac{3}{4}$
28) $5\frac{3}{5}$

Comparing Decimals

1) $0.632 > 0.631$
2) $0.75 < 1$
3) $3.91 < 4.91$
4) $3.2 > 3.1$
5) $2.8 < 2.801$
6) $0.47 < 0.74$
7) $14.9 > 1.49$
8) $0.707 > 0.0707$
9) $1.01 > 0.999$
10) $3.655 < 6.6555$
11) $15.4 > 14.5$
12) $0.909 < 0.99$
13) $3.3 < 3.33$
14) $0.304 = 0.304$
15) $4.0001 < 4.001$
16) $3.003 < 3.3$
17) $2.85 < 2.88$
18) $0.98 > 0.908$
19) $2.031 > 2.0031$
20) $5.97 > 5.79$
21) $6.302 > 6.203$
22) $0.075 < 0.57$
23) $1.04 < 1.0401$
24) $9.101 > 9.011$
25) $0.259, 2.54, 3.05, 5.46, 8.32$
26) sparkling water

Rounding Decimals

1) 3.1
2) 0.36
3) 0.5
4) 7
5) 1.72
6) 3.28
7) 3.3
8) 11
9) 45
10) 7.1
11) 12.65
12) 7
13) 5.9
14) 101
15) 2
16) 0.4
17) 100
18) 8
19) 22.7
20) 30.3
21) 47.9
22) 82.9
23) 16.2
24) 71.8
25) 5.44
26) 12.91
27) 26.19
28) 48.62
29) 91.45
30) 29.35
31) 23
32) 9
33) 14
34) 8
35) 4
36) 57
37) 14
38) 13
39) 14
40) 156
41) 7
42) 12
43) 51
44) 32
45) 101

Adding and Subtracting Decimals

1) 7.43
2) 21.13
3) 30.79
4) 29.61
5) 27.97
6) 114.1
7) 26.91
8) 23.76
9) 144.01
10) 23.61
11) -4.98
12) 31.54
13) 11.14
14) 7.1
15) 4.72
16) 4.41
17) 1.75
18) 3.65
19) 1.35
20) 1.19

21) $679.14

Multiplying and Dividing Decimals

1) 21.45
2) 34.78
3) 1.8315
4) 39.325
5) 124.125
6) 146.633
7) 704.77
8) 219.96
9) 20.2875
10) 78.4
11) 42.4
12) 3.14
13) 8.2
14) 3.38
15) 3.9
16) 1.71
17) 0.99
18) 3.63
19) 2.81
20) 1.25
21) 24.175

Converting Between Fractions, Decimals and Mixed Numbers

1) 0.7
2) 0.6
3) 0.33
4) 0.375
5) 0.25
6) 3
7) 3.5
8) 5
9) 6.6
10) Darcy, 0.833 ...
11) Sierra(0.5), Jacob(0.6), Sara(0.75)
12) James(0.5), Garret(0.833), Massey(1.33)
13) $\frac{7}{10}$
14) $\frac{1}{4}$
15) $4\frac{3}{10}$
16) $9\frac{25}{100}$
17) $4\frac{7}{10}$
18) $15\frac{5}{10}$

Factoring Numbers

1) 1,5,7,35
2) 1,17
3) 1,2,3,6,7,14,21,42
4) 1,2,3,4,6,8,12,24
5) 1,3,11,33
6) 1,2,11,22
7) 1,3,13,39
8) 1,3,17,51
9) 1,2,17,34
10) 1,2,3,6,9,18
11) 1,3,23,69
12) 1,2,29,58
13) 1,2,4,19,38,76
14) 1,2,3,4,6,8,12,16,24,48
15) 1,2,7,14
16) 1,2,3,4,6,12
17) 1,2,3,4,6,9,18
18) 1,2,4,5,10,20
19) 1,2,3,4,6,8,12,24
20) 1,2,3,4,6,8,12,18,24,36,72
21) 1,5,17,85

22) 1,2,19,38
23) 1,5,7,35
24) 1,2,4,8
25) 1,5
26) 1,2,4
27) 1,2,3,4,6,9,12,18,36
28) 1,2,3,6,7,14,21,42
29) 1,2,4,7,8,14,28,56
30) 1,3,7,9,21,63
31) 1,2,4,5,8,10,16,20,40,80
32) 1,5,19,95
33) 1,2,3,6,17,34,51,102
34) $2 \times 3 \times 5$
35) $2 \times 2 \times 2 \times 7$
36) $2 \times 3 \times 13$
37) 5×5
38) 2×23
39) $2 \times 2 \times 7$
40) $3 \times 3 \times 7$
41) $2 \times 2 \times 13$
42) $2 \times 3 \times 3$
43) $2 \times 2 \times 2 \times 2 \times 3$
44) 2×29
45) $2 \times 2 \times 3 \times 3$
46) $2 \times 2 \times 31$
47) $2 \times 5 \times 9$
48) 3×23
49) $2 \times 2 \times 2 \times 3 \times 3$
50) 5×17
51) 3×7
52) $2 \times 3 \times 4$
53) $2 \times 3 \times 3$
54) $2 \times 2 \times 2 \times 3$
55) 5×11
56) $3 \times 5 \times 5$
57) 3×3
58) 2×5
59) 3×5
60) 2×7
61) $2 \times 3 \times 3$
62) 19
63) $2 \times 2 \times 5$

Greatest Common Factor

1) 12
2) 2
3) 5
4) 6
5) 3
6) 14
7) 5
8) 5
9) 10
10) 4
11) 17
12) 2
13) 3
14) 5
15) 4
16) 8
17) 8
18) 3
19) 3

Least Common Multiple

1) 50
2) 36
3) 40
4) 36
5) 96
6) 70
7) 56
8) 969
9) 60
10) 420
11) 75
12) 50
13) 63
14) 78
15) 110
16) 48
17) 63
18) 300
19) 70
20) 72
21) 90
22) 36
23) 72
24) 99

Chapter 2:

Real Numbers and Integers

Topics that you'll learn in this part:

- ✓ Adding and Subtracting Integers
- ✓ Multiplying and Dividing Integers
- ✓ Ordering Integers and Numbers
- ✓ Arrange and Order, Comparing Integers
- ✓ Order of Operations
- ✓ Mixed Integer Computations
- ✓ Integers and Absolute Value

Adding and Subtracting Integers

✏ Find the sum.

1) $(-37) + (-8) =$

2) $8 + (-17) =$

3) $(-53) + (-7) =$

4) $(-41) + (23) =$

5) $(-14) + (-5) =$

6) $(-72) + (-30) + 2 =$

7) $4 + (-40) + (-15) + (-21) =$

8) $91 + (-143) + (-45) =$

9) $(-33) + (-18) =$

10) $(-14) + (58 - 44) =$

✏ Find the difference.

11) $(-34) - (-28) - 4 =$

12) $54 - (-12) =$

13) $(-35) - (-5) =$

14) $(-51) - (-34) =$

15) $(-45) - (-30) =$

16) $(-15) - (-10) =$

17) $(-30) - (-14) - (-17) =$

18) $(-10) - (-10) - (-3) =$

19) $(-45) - (-17) =$

20) $(-7) - (-34) - 17 =$

✏ Solve.

21) The leaderboard at the Stamford Golf Tournament shows that Nancy's score is 5 and Doug's score is (-1). How many more strokes did Nancy take than Doug?

22) Bridget carefully tracks her money. Her records indicate she spent $300 on a hammock and deposited $1,000 she made from an online auction. Which integer represents the change in how much money Bridget had?

Multiplying and Dividing Integers

✏ Find each product.

1) $(-11) \times (-5) =$
2) $34 \times (-2) =$
3) $(-4) \times 5 =$
4) $7 \times (-10) =$
5) $(-11) \times (-2) \times 2 =$

6) $6 \times (-15) =$
7) $14 \times (-14) =$
8) $(-13) \times (-10) =$
9) $(-14) \times (-4) \times (-5) =$
10) $14 \times (-5) =$

✏ Find each quotient.

11) $210 \div (-14) =$
12) $(-208) \div (-13) =$
13) $(108) \div (-9) =$
14) $(-161) \div (-23) =$

15) $84 \div (-14) =$
16) $(-484) \div (-22) =$
17) $(-162) \div (-18) =$
18) $198 \div 6 =$

✏ Solve.

19) Adam is scuba diving. He descends 5 feet. He descends the same distance 4 more times. What integer represents Adam's new DISTANCE from sea level?

20) The price of jeans was reduced $6 per week for 7 weeks. By how much did the price of the jeans change over the 7 weeks?

21) Yesterday's low temperature was (-2°C). Today's low temperature is 3 times as low as yesterday's low temperature. What is today's low temperature?

Ordering Integers and Numbers

Order each set of integers from least to greatest.

1) $36, -10, 0, 17, 2, -12$ — $-12, -10, 0, 2, 17, 36$
2) $43, 10, 21, -30, -1, -12, 2$ — $-30, -12, -1, 2, 10, 21, 43$
3) $45, -10, 14, -14, 0, -3$ — $-14, -10, -3, 0, 14, 45$
4) $-100, 0, 100, 10$ — $-100, 0, 10, 100$
5) $56, -2, -3, -50$ — $-50, -3, -2, 56$
6) $18, -6, 15, -1, -10$ — $-10, -6, -1, 15, 18$
7) $-20, 12, 0, 15, -30, -2$ — $-30, -20, -2, 0, 12, 15$
8) $50, 12, -52, -12, -3$ — $-52, -12, -3, 12, 50$
9) $-9, -1, 0, 2, 3, -6$ — $-9, -6, -1, 0, 2, 3$
10) $12, 21, -14, 8, -9, 10$ — $-14, -9, 8, 10, 12, 21$

Order each set of integers from greatest to least.

11) $-99, 7, 10, 0$ — $10, 7, 0, -99$
12) $5, -4, -2, 0, 10$ — $10, 5, 0, -2, -4$
13) $-30, -100, 33, -33$ — $33, -30, -33, -100$
14) $-81, 10, 71, 23, 51, 12, -3$ — $71, 51, 23, 12, 10, -3, -81$
15) $-3, -2, 6, -32, 5, 12$ — $12, 6, 5, -2, -3, -32$
16) $13, -1, 1, 0, -13$ — $13, 1, 0, -1, -13$
17) $79, 0, 12, -100$ — $79, 12, 0, -100$
18) $99, -1, 23, -3, 0$ — $99, 23, 0, -1, -3$
19) $44, -6, -100, 19$ — $44, 19, -6, -100$

Arrange, Order, and Comparing Integers

✍ Arrange these integers in descending order.

1) $34, 15, -3, -4$ ___, ___, ___, ___, ___, ___

2) $17, -10, 0, -14$ ___, ___, ___, ___, ___, ___

3) $15, -78, -15, -1$ ___, ___, ___, ___, ___, ___

4) $35, -17, 12, -45$ ___, ___, ___, ___, ___, ___

5) $-30, -10, 71, -15, -14$ ___, ___, ___, ___, ___, ___

6) $62, -20, 12, 15, 14, 0$ ___, ___, ___, ___, ___, ___

7) $-369, 12, -1, 1, 0, -95$ ___, ___, ___, ___, ___, ___

8) $62, -41, -1, 3, 14, 7, -7$ ___, ___, ___, ___, ___, ___

9) $36, -100, 100, -5, 5$ ___, ___, ___, ___, ___, ___

10) $99, -87, 56, -45, -110$ ___, ___, ___, ___, ___, ___

✍ Compare. Use >, =, <

11) $-15 \square 12 =$

12) $-33 \square -16 =$

13) $-65 \square 0 =$

14) $30 \quad -35 =$

15) $96 \quad -96 =$

16) $-568 \square -658 =$

17) $-321 \square -321 =$

18) $545 \square -545 =$

19) $-1000 \square -965 =$

20) $-25 \square -656 =$

21) $-89 \square -100 =$

22) $0 \square -1 =$

23) $10 \square -11 =$

24) $79 \square -100 =$

25) $95 \quad 89 =$

26) $0.2 \square -0.2 =$

27) $12 \square -15 =$

28) $9 \square -9 =$

Order of Operations

✎Evaluate each expression.

1) $15 + \left(\frac{66}{12-((-5)\times 2)}\right) =$

2) $\frac{(-55)}{5} + 10 =$

3) $(-13) + (6 \div 2) =$

4) $\frac{45}{(-15)} + (3 \times 2) =$

5) $(-40) + (6 \times (-7)) =$

6) $(5 \times (-7)) - \frac{40}{8} =$

7) $(-78) - \frac{28}{(-2)} =$

8) $\frac{(-80)}{(-4)} =$

9) $15 - (3 \times (-7)) =$

10) $45 + (3 \times (-15)) =$

11) $45 - \left(\frac{3 \times 10}{4+(-2)}\right) =$

12) $(-48) - \frac{8}{2} =$

13) $\frac{(-35)}{7} + 3 =$

14) $\left(\frac{70}{(-14)}\right) - (2 \times 3) =$

15) $36 + ((-2) \times 3) =$

16) $15 - \frac{30}{(-6)} =$

17) $78 + (-13) =$

18) $6 + \frac{(-51)}{(-17)} =$

19) $5 - \frac{25}{-10} =$

20) $\frac{3(15-8)}{7} + 5 =$

21) $\left(\frac{25-13}{2(3)} - 7\right) \times 2 =$

22) $5 \times \left(\frac{55}{11}\right) + 8 =$

23) $(-12) \times \frac{48}{6} + 12 =$

24) $\frac{96}{-12} - 6 =$

25) $\frac{(33 \times 2)}{(-6) \times 11} + 1 =$

26) $12 \times \left(\frac{45}{(-5)}\right) =$

✎Solve.

27) Sylvia bought 6 bananas for 60 cents each and 1 apple for 90 cents. Write a numerical expression to represent this situation and then find the total cost in dollar.

Mixed Integer Computations

✎ Compute.

1) $(-70) \div \left(\frac{20}{4}\right) =$

2) $(-21) \times \frac{(-3)}{7} =$

3) $(-10) \times \left(-\frac{14}{5}\right) =$

4) $\left(\frac{3+(-13)}{2}\right) \div 5 =$

5) $18 \times \frac{24}{(-18)} =$

6) $(-90) \div \frac{(-45)}{14} =$

7) $\frac{\left(-\frac{48}{4}\right)}{\left(\frac{60}{30}\right)} \times \frac{25}{(-5)} =$

8) $\left(2 \times \frac{24}{(-4)}\right) \div (-6) =$

9) $78 \div (-6) =$

10) $\frac{(-27)}{3} \times \left(-\frac{14}{7}\right) =$

11) $5 \times \left((-4) + \frac{15}{5}\right) =$

12) $\frac{(-48)}{4} \div (-2) =$

13) $\frac{(-36)}{12} \times (-2) =$

14) $(-10) \times (9) =$

15) $\frac{30}{(-6)} \times \frac{(-45)}{(-15)} =$

16) $(-100) \div \left(\frac{(-100)}{45}\right) =$

17) $(-80) \div (-20) =$

18) $(-6) \times (-11) =$

19) $(-3) \times \frac{(14 \times (-3))}{42} =$

20) $2 \times \left(-\frac{56}{8}\right) =$

21) $\frac{5}{2} \div \frac{10}{6} =$

22) $(-24) \div \frac{24}{10} + 5 =$

23) $\frac{25}{(-5)} \div \frac{1}{5} =$

24) $\frac{72 \div 12}{35 \div 5} \times 7 =$

25) $\frac{3(50 \div 5)}{30} \div \frac{10}{45} =$

26) $\frac{25}{5} \div \frac{25}{125} =$

27) $\frac{56}{7} \times \left(3 - \frac{42}{7}\right) =$

28) $3 \times \frac{24}{6} + 5 =$

29) $\frac{3}{10} \times \frac{-10}{7} =$

30) $\frac{(-25)}{11} \div \frac{40}{11} =$

31) $45 \div \left(\frac{4}{3} \div \frac{8}{6}\right) =$

32) $(-3) \div \left(\frac{(-20)}{66}\right) =$

33) $\left(\frac{45}{20}\right) \times \left(\frac{15}{20} \div \frac{10}{12}\right) =$

34) $\frac{1}{2} \times \left(3 - \frac{1}{3}\right) =$

Integers and Absolute Value

✎ *Write absolute value of each number.*

1) -78
2) -1
3) -65
4) 56
5) -98
6) -11
7) -21
8) 3
9) 2
10) -42
11) -61
12) -10
13) 36
14) -13
15) -33
16) -22
17) -19
18) 17
19) -14
20) -41
21) -98

✎ *Evaluate.*

22) $|-22| - |12| =$

23) $12 + |-15 - 10| - |-3| =$

24) $|-16| - 40 + 30 =$

25) $|-113| - |(-35) + 30| =$

26) $|12 - 4| + 6 - |-9| =$

27) $|-52| + |-10| =$

28) $|-2 + 8| + |7 - 7| =$

29) $|-12| + |-11| =$

✎ *Solve.*

30) You have money in your wallet, but you don't know the exact amount. When a friend asks you, you say that you have 50 dollars give or take 15. Use an absolute value equation to find least and biggest amount of money in your pocket?

31) The ideal selling price of a Toyota is 25000. The dealer allows this price to vary 5%. What is the lowest price this dealer can sell this Toyota?

Answers of Worksheets

Adding and Subtracting Integers

1) −45
2) −9
3) −60
4) −18
5) −19
6) −100
7) −72
8) −97
9) −51
10) 0
11) −10
12) 66
13) −30
14) −17
15) −15
16) −5
17) 1
18) 3
19) −28
20) 10
21) 6
22) 700

Multiplying and Dividing Integers

1) 55
2) −68
3) −20
4) −70
5) 44
6) −90
7) −196
8) 130
9) −280
10) −70
11) −15
12) 16
13) −12
14) 7
15) −6
16) 22
17) 9
18) 33
19) −25
20) −42
21) −6

Ordering Integers and Numbers

1) −12, −10, 0, 2, 17, 36
2) −30, −12, −1, 2, 10, 21, 43
3) −14, −10, −3, 0, 14, 45
4) −100, 0, 10, 100
5) −50, −3, −2, 56
6) −10, −6, −1, 15, 18
7) −30, −20, −2, 0, 12, 15
8) −52, −12, −3, 12, 50
9) −9, −6, −1, 0, 2, 3
10) −14, −9, 10, 12, 21
11) 10, 7, 0, −99
12) 10.5, 0, −2, −4
13) 33, −30, −33, −100
14) 71, 51, 23, 12, 10, −3, −81
15) 12, 6, 5, −2, −3, −32
16) 13, 1, 0, −1, −13
17) 79, 12, 0, −100
18) 99, 23, 0, −1, −3
19) 44, 19, −6, −100

Arrange and Order, Comparing Integers

1) 34, 15, −3, −4
2) 17, 0, −10, −14
3) 15, −1, −15, −78
4) 35, 12, −17, −45
5) 71, −10, −14, −15, −30
6) 62, 15, 14, 12, 0, −20
7) 12, 1, 0, −1, −95, −369
8) 62, 14, 7, 3, −1, −7, −41
9) 100, 36, 5, −5, −100
10) 99, 56, −45, −87, −110
11) <
12) <

13) <   19) <   25) >
14) >   20) >   26) >
15) >   21) >   27) >
16) >   22) >   28) >
17) =   23) >
18) >   24) >

Order of Operations

1) 18   8) 20   15) 30   22) 33
2) −1   9) 36   16) 20   23) −84
3) −10   10) 0   17) 65   24) −14
4) 3   11) 30   18) 9   25) 0
5) −82   12) −52   19) 7.5   26) −108
6) −40   13) −2   20) 8   27) $4.5
7) −64   14) −11   21) −10

Mixed Integer Computations

1) −14   10) 18   19) 3   28) 17
2) 9   11) −5   20) −14   29) $-\frac{3}{7}$
3) 28   12) 6   21) $\frac{3}{2}$ 30) $\frac{(-25)}{40}$
4) −1   13) 6   22) −5   31) 45
5) −24   14) −90   23) −25   32) $\frac{99}{10}$
6) 28   15) −15   24) 6   33) $\frac{81}{40}$
7) 30   16) 45   25) 4.5   34) $\frac{8}{6}$
8) 2   17) 4   26) 25
9) −13   18) 66   27) −24

Integers and Absolute Value

1) 78   3) 65   5) 98   7) 21
2) 1   4) 56   6) 11   8) 3

9) 2	15) 33	21) 98	27) 62
10) 42	16) 22	22) 10	28) 6
11) 61	17) 19	23) 34	29) 23
12) 10	18) 17	24) 6	30) 35,65
13) 36	19) 14	25) 108	31) 23750
14) 13	20) 41	26) 5	

Chapter 3:

Proportions and Ratios

Math Topics that you'll learn in this part:

- ✓ Writing Ratios
- ✓ Simplifying Ratios
- ✓ Proportional Ratios
- ✓ Create a Proportion
- ✓ Similar Figures
- ✓ Similar Figure Word Problems
- ✓ Ratio and Rates Word Problems

Writing Ratios

✏️ ***Express each ratio as a rate and unite rate.***

1) 150 miles on 5 gallons of gas.

2) 20 dollars for 4 books.

3) 100 miles on 8 gallons of gas

4) 30 inches of snow in 10 hours

✏️ ***Express each ratio as a fraction in the simplest form.***

5) 6 feet out of 24 feet

6) 12 cakes out of 24 cakes

7) 7 dimes of 35 dimes

8) 16 dimes out of 48 coins

9) 21 cups to 56 cups

10) 36 gallons to 85 gallons

11) 18 miles out of 42 miles

12) 23 blue cars out of 46 cars

✏️ ***Solve.***

13) In a telephone poll, 10 people said they like shopping and 20 people said they do not like shopping. What is the ratio of the number of people who do not like shopping to the number of people who like shopping?

14) There are 30 pink beads and 6 purple beads on Maria's necklace. What is the ratio of the number of pink beads to the number of purple beads?

15) 40 of the tables at Gary's Italian Restaurant are full and the other 8 tables are empty. What is the ratio of the number of full tables to the number of empty tables?

Simplifying Ratios

✏️ *Reduce each ratio.*

1) 42 : 70
2) 18 : 54
3) 15 : 45
4) 21 : 45
5) 10 : 60
6) 46 : 92
7) 50 : 20
8) 12 : 40

9) 90 : 45
10) 20 : 85
11) 28 : 40
12) 12 : 72
13) 24 : 12
14) 55 : 25
15) 39 : 13
16) 14 : 77

17) 18 : 66
18) 8 : 32
19) 45 : 100
20) 6 : 30
21) 17 : 85
22) 36 : 90
23) 15 : 80
24) 40 : 100

✏️ *Each pair of figures is similar. Find the missing side.*

25)
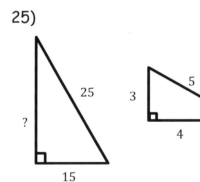

26)
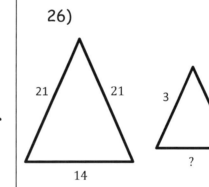

27)
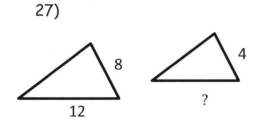

28)
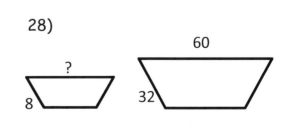

Create a Proportion

✎ *Create proportion from the given set of numbers.*

1) 1, 15, 5, 3
2) 12, 36, 4, 12
3) 32, 8, 8, 2
4) 17, 5, 51, 15

5) 9, 7, 54, 42
6) 56, 5, 8, 35
7) 3, 5, 55, 33
8) 12, 12, 3, 48

✎ *Solve.*

9) In a party, 10 soft drinks are required for every 12 guests. If there are 252 guests, how many soft drinks is required?

10) Mika can eat 21 hot dogs in 6 minutes. She wants to know how many minutes (m) it would take her to eat 35 hot dogs if she can keep up the same pace.

11) Mandy works construction. She knows that a 5-meter-long metal bar has a mass of 40kg. Mandy wants to figure out the mass (w) of a bar made of the same metal that is 3 meters long and the same thickness. What is the mass of the shorter bar?

12) Kwesi is putting on sunscreen. He uses 3ml to cover 45cm² of his skin. He wants to know how many milliliters of sunscreen (g) he needs to cover 240cm² of his skin. He assumes the relationship between milliliters of sunscreen and area is proportional. How many milliliters of sunscreen does Kwesi need to cover 240cm² of his skin?

Similar Figures

✎ *Each pair of figures is similar. Find the missing side.*

1)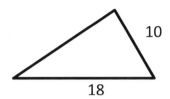

2)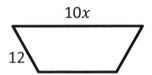

3)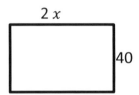

4)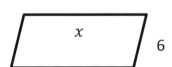

5)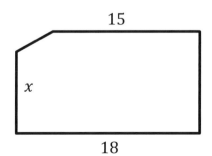

Simple Interest

✎ Use simple interest to find the ending balance.

1) $1,200 at 3% for 5 years.

2) $5,650 at 6% for 4 months.

3) $1600 at 8% for 8 years

4) $12,000 at 6.5% for 6 years.

5) $3200 at 4% for 7 years.

6) $32,500 at 8% for 6 years.

✎ Solve.

7) $300 interest is earned on a principal of $2000 at a simple interest rate of 5% interest per year. For how many years was the principal invested?

8) A new car, valued at $30,000, depreciates at 8% per year from original price. Find the value of the car 2 years after purchase.

9) Sara puts $4,000 into an investment yielding 7% annual simple interest; she left the money in for four years. How much interest does Sara get at the end of those four years?

10) You want to save $1,200 to buy your first self-driving magic carpet. You deposit $8,000 in a bank at an interest rate of 5% per annum. How many years do you have to wait before you can buy your magic carpet?

11) Aladdin has 12 extra gold coins in his magic bag. The Genie tells him that for every 100 gold coins he has in his magic bag, he will get 25 extra gold coins every year. How many years later will Aladdin have 21 gold coins in his bag?

Ratio and Rates Word Problems

✏️ *Solve.*

1) In a party, 10 soft drinks are required for every 14 guests. If there are 266 guests, how many soft drinks is required?

2) In Jack's class, 12 of the students are tall and 8 are short. In Michael's class 26 students are tall and 14 students are short. Which class has a higher ratio of tall to short students?

3) Are these ratios equivalent? 13 cards to 78 animals 15 marbles to 90 marbles.

4) The price of 4 apples at the Quick Market is $3. The price of 7 of the same apples at Walmart is $5.50. Which place is the better buy?

5) The bakers at a Bakery can make 200 bagels in 5 hours. How many bagels can they bake in 15 hours? What is that rate per hour?

6) You can buy 8 cans of green beans at a supermarket for $5.20. How much does it cost to buy 45 cans of green beans?

7) Finley makes 11 batch of her favorite shade of orange paint by mixing 55 liters of yellow paint with 33 liters of red paint. How many batches of orange paint can Finley make if she has 15 liters of red paint?

8) Quinn is playing video games at a virtual reality game room. The game room charges 20 dollars for every 30 minutes of play time. How much does Quinn need to pay for 150 minutes of play time.

Answers of Worksheets

Writing Ratios

1) $\frac{150\ miles}{5\ gallons}$, 30 miles per gallon

2) $\frac{20\ dollars}{4\ books}$, 5.00 dollars per book

3) $\frac{100\ miles}{8\ gallons}$, 12.5 miles per gallon

4) $\frac{30"\ of\ snow}{10\ hours}$, 3 inches of snow per hour

5) $\frac{1}{4}$ 7) $\frac{1}{5}$ 9) $\frac{3}{8}$ 11) $\frac{3}{7}$ 14) 5

6) $\frac{1}{2}$ 8) $\frac{1}{3}$ 10) $\frac{36}{85}$ 12) $\frac{1}{2}$ 15) 5

13) 2

Simplifying Ratios

1) 3:5   7) 5:2   13) 2:1   19) 9:20   25) 20

2) 1:3   8) 3:10   14) 11:5   20) 1:5   26) 2

3) 1:3   9) 2:1   15) 3:1   21) 1:6   27) 6

4) 7:15   10) 4:17   16) 2:11   22) 2:5   28) 15

5) 1:6   11) 7:10   17) 3:11   23) 3:16

6) 1:2   12) 1:6   18) 1:4   24) 2:5

Create a Proportion

1) 1:3 = 5:15   5) 7:42 = 9:54   9) 210

2) 12:36 = 4:12   6) 5:35 = 8:56   10) 10

3) 2:8 = 8:32   7) 3:33 = 5:55   11) 24

4) 5:15 = 17:51   8) 3:12 = 12:48   12) 16

Similar Figures

1) 9   2) 3   3) 24   4) 15   5) 6

Simple Interest

1) $1380.00   4) $16,680.00   7) 3 years   9) $1,120

2) $7,006.00   5) $4,096.00   8) $25,200   10) 3 years

3) $2,624.00   6) $48,100.00

11) %5

Ratio and Rates Word Problems

1) 190

2) The ratio for Michael's class is higher and equal to 13 to 7.

3) Yes! Both ratios are 1 to 6

4) The price at the Quick Market is a better buy.

5) 600, the rate is 40 per hour.

6) $29.25

7) 5

8) 100

Chapter 4:

Percent

Math Topics that you'll learn in this part:

- ✓ Percentage Calculations
- ✓ Converting Between Percent, Fractions, and Decimals
- ✓ Percent Problems
- ✓ Find What Percentage a Number Is of Another
- ✓ Find a Percentage of a Given Number
- ✓ Percent of Increase and Decrease
- ✓ Markup, Discount, and Tax

Percentage Calculations

✎ Calculate the percentages.

1) 10% of 50 =
2) 15% of 80 =
3) 50% of 26 =
4) 30% of 20 =
5) 45% of 100 =
6) 25% of 100 =
7) 70% of 30 =
8) 38% of 50 =
9) 20% of 50 =
10) 75% of 100 =
11) 65% of 80 =
12) 30% of 50 =
13) 20% of 0 =
14) 84% of 200 =
15) 12% of 100 =
16) 40% of 300 =
17) 30% of 60 =
18) 50% of 90 =
19) 30% of 45 =
20) 60% of 150 =

✎ Solve.

21) A test has 20 questions. If peter gets 80% correct, how many questions did peter missed?

22) In a school, 25 % of the teachers teach basic math. If there are 50 basic math teachers, how many teachers are there in the school?

23) 24 students in a class took an algebra test. If 18 students passed the test, what percent do not pass?

24) Yesterday, there were 20 problems assigned for math homework. Lucy got 18 out of 20 problems correct. What percentage did Lucy get correct?

25) Megan's Tea Shop has caffeinated tea and decaffeinated tea. The tea shop served 10 teas in all, 7 of which were caffeinated. What percentage of the teas were caffeinated?

Converting Between Percent, Fractions, and Decimals

✍ Converting fractions to decimals.

1) $\frac{35}{100}$
2) $\frac{32}{100}$
3) $\frac{56}{100}$
4) $\frac{46}{100}$
5) $\frac{72}{100}$
6) $\frac{21}{100}$
7) $\frac{91}{100}$
8) $\frac{36}{100}$
9) $\frac{98}{100}$

✍ Write each decimal as a percent.

10) 0.72
11) 0.962
12) 0.54
13) 0.42
14) 0.83
15) 0.452
16) 1.3
17) 0.035
18) 3.12

✍ How many percentages have the sizes changed?

19)

20)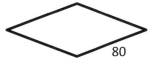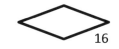

21)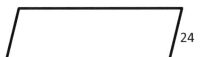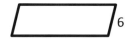

22)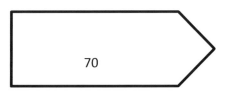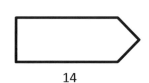

Percent Problems

✒ Solve each problem.

1) 60 is 120% of what?

2) 40% of what number is 50?

3) 27% of 142 is what number?

4) What percent of 125 is 30?

5) 30 is what percent of 120?

6) 44 is 25% of what?

7) 33 is 30% of what?

8) 52% of 150 is what?

9) 15 is what percent of 300?

10) What is 35% of 52 m?

11) What is 45% of 120 inches?

12) 18 inches is 40% of what?

✒ Solve.

13) Liam scored 21 out of 33 marks in Algebra, 37 out of 46 marks in science and 75 out of 95 marks in mathematics. In which subject his percentage of marks in best?

14) Ella require 50% to pass. If she gets 160 marks and falls short by 40 marks, what were the maximum marks she could have got?

15) There are 60 employees in a company. On a certain day, 36 were present. What percent showed up for work?

16) A metal bar weighs 24 ounces. 15% of the bar is gold. How many ounces of gold are in the bar?

17) A crew is made up of 12 women; the rest are men. If 20% of the crew are women, how many people are in the crew?

Find What Percentage a Number Is of Another

✏️ *Find the percentage of the numbers.*

1) 10 is what percent of 40?

2) 12 is what percent of 240?

3) 151.2 is what percent of 270?

4) 14 is what percent of 112?

5) 6 is what percent of 30?

6) 17 is what percent of 85?

7) 39.6 is what percent of 88?

8) 420 is what percent of 350?

9) 13 is what percent of 104?

10) 9 is what percent of 225?

11) 75 is what percent of 50?

12) 11 is what percent of 55?

13) 300 is what percent of 1250?

14) 7.5 is what percent of 60?

15) 352 is what percent of 220?

16) 504 is what percent of 252?

17) 52 is what percent of 260?

18) 45 is what percent of 360?

✏️ *Solve*

19) Challenger Elementary School has 800 students. Every Wednesday, 12% of the students stay after school for Chess Club. How many students attend Chess Club on Wednesdays?

20) Anastasia is grocery shopping with her father and wonders how much shopping is left to do. "We already have 60% of the items on our list," her father says. Anastasia sees 12 items in the cart. How many grocery items are on the list?

21) A gumball machine contains 23 green gumballs, 52 red gumballs, 34 blue gumballs, 61 yellow gumballs, and 30 pink gumballs. What percentage of the gumballs are red?

Find a Percentage of a Given Number

✎ Find a Percentage of a Given Number.

1) 25% of 50 =
2) 40% of 90 =
3) 11% of 69 =
4) 12% of 60 =
5) 45% of 66 =
6) 38% of 55 =
7) 65% of 30 =
8) 15% of 40 =
9) 5% of 80 =
10) 20% of 80 =
11) 80% 0f 80 =
12) 25% of 36 =
13) 70% of 40 =
14) 45% of 60 =
15) 45% of 90 =
16) 40% of 120 =
17) 30% of 9 =
18) 65% of 200 =
19) 66% of 10 =
20) 36% of 50 =
21) 96% of 80 =

✎ Solve.

22) Luke and Matthew ran a lemonade stand on Saturday. They agreed that Matthew would get 60% of the profit because the lemonade stand was his idea. They made a profit of $25. How much money did Matthew make?

23) Mrs. Conley asks her class what kind of party they want to have to celebrate their excellent behavior. Out of all the students in the class, 5 want an ice cream party, 7 want a movie party, 10 want a costume party, and the rest are undecided. If 20% want an ice cream party, how many students are in the class?

24) There are 25 students in Ms. Nguyen's second-grade class. In the class election, 4 students voted for Benjamin, 12 voted for Sahil, and 9 voted for Maria. What percentage of the class voted for Maria?

Percent of Increase and Decrease

✎ *Find each percent change to the nearest percent. Increase or decrease.*

1) From 25 grams to 110 grams.
2) From 200 m to 50 m
3) From $520 to $102
4) From 256 ft. to 70 ft.
5) From 526 ft. to 800 ft.
6) From 25 inches to 125 inches
7) From 33 ft. to 163 ft.
8) From 536 miles to 76 miles

✎ *Solve.*

9) The population of a place in a particular year increased by 10%. Next year it decreased by 15%. Find the net increase or decrease percent in the initial population.

10) While measuring a line segment of length 5cm, it was measured 5.2cm by mistake. Find the percentage error in measuring the line segment.

11) A number is increased by 40% and then decreased by 40%. Find the net increase or decrease per cent.

12) The price of wheat increased by 10%. By how much per cent should mother reduce her consumption in the house so that her expenditure on wheat does not increase?

13) The football team at Riverside College plays in an old stadium that seats 31,780 people. This stadium will be demolished and a new one built that can hold 35% more fans. What will be the seating capacity of the new, bigger stadium?

Markup, Discount, and Tax

✎ Find the selling price of each item.

1) Cost of a pen: $4.5, markup:.25%, discount: 20%, tax:5%

2) Cost of a puppy: $210, markup:20%, discount:15%

3) Cost of a shirt: $18.00, markup:22%, discount:20%

4) Cost of an oil change: $28.5, markup:55%

5) Cost of computer: $1,690.00, markup:15%

✎ Solve.

6) Vanessa earns a base salary of $400.00 with an additional %5, percent commission on everything she sells. Vanessa sold $1650.00-dollar worth of items last week. What was Vanessa's total pay last week?

7) The pie store is having a %20 percent off sale on all its pies. If the pie you want regularly costs $18, how much would you save with the discount?

8) Zoe paid $18.60 in sales tax for purchasing a table. The sales tax rate is 12%, What was the price of Zoe's table before sales tax?

9) Daniel works at a nearby electronics store. He makes a commission of 15% on everything he sells. If he sells a laptop for $293.00, how much money does Daniel make in commission?

Answers of Worksheets

Percentage Calculations

1) 5
2) 12
3) 13
4) 6
5) 45
6) 25
7) 21
8) 19
9) 10
10) 75
11) 52
12) 15
13) 0
14) 168
15) 12
16) 120
17) 18
18) 45
19) 13.5
20) 90
21) 4
22) 200
23) 25%
24) 90%
25) 70%

Converting Between Percent, Fractions, and Decimals

1) 0.35
2) 0.32
3) 0.56
4) 0.46
5) 0.72
6) 0.21
7) 0.91
8) 0.36
9) 0.98
10) 72%
11) 96.2%
12) 54%
13) 42%
14) 83%
15) 45.2 %
16) 130%
17) 3.5%
18) 312%
19) 75%
20) 20%
21) 25%
22) 20%

Percent Problems

1) 50
2) 125
3) 38.34
4) 24%
5) 25%
6) 176
7) 110
8) 78
9) 5%
10) 18.2 m
11) 54 inches
12) 45 inches
13) science, 80%
14) 400
15) 60%
16) 3.6 ounces
17) 60

Find What Percentage a Number Is of Another

1) 25%
2) 5%
3) 56%
4) 12.5%
5) 20%
6) 20%
7) 45%
8) 120%
9) 12.5%
10) 4%
11) 150%
12) 20%
13) 24%
14) 12.5%
15) 160%
16) 200%
17) 20%
18) 12.5%
19) 96
20) 20
21) 26%

Find a Percentage of a Given Number

1) 12.5	6) 20.9	11) 64	16) 48	21) 76.8
2) 36	7) 19.5	12) 9	17) 2.7	22) 15
3) 7.59	8) 6	13) 28	18) 130	23) 25
4) 7.2	9) 4	14) 27	19) 6.6	24) 36
5) 29.7	10) 16	15) 40.5	20) 18	

Percent of Increase and Decrease

1) 440% increase
2) 75% decrease
3) 80.4% decrease
4) 73% decrease
5) 52% increase
6) 500% increase
7) 493% increase
8) 86% decrease
9) 6.5% decrease
10) 4%
11) 16% decrease
12) $9\frac{1}{10}\%$
13) 42,903

Markup, Discount, and Tax

1) $4.725
2) $214.2
3) $17.568
4) $44.175
5) $1943.5
6) $482.50
7) 3.6
8) 16.61
9) 43.95

Section 2: Algebra

- *Algebraic Expressions*
- *Equations and Inequalities*
- *Systems of Equations*
- *Linear Functions*
- *Monomials and Polynomials*
- *Exponents and Radicals*

Chapter 5:

Algebraic Expressions

Topics that you'll learn in this part:

- ✓ Expressions and Variables
- ✓ Simplifying Variable Expressions
- ✓ Simplifying Polynomial Expressions
- ✓ The Distributive Property
- ✓ Evaluating One Variable
- ✓ Evaluating Two Variables
- ✓ Combining like Terms

Expressions and Variables

✎ Simplify each expression.

1) $3x + 2x$,
   Use $x = 4$

2) $3(-3x + 9) + 3x$,
   Use $x = 2$

3) $(2x + 5) + (-2x)$,
   Use $x = 4$

4) $(6x - 3)(2x + 1)$,
   Use $x = 2$

5) $(2x + 5) + (3y - 3)$,
   Use $x = 2, y = -2$

6) $(5x - 1)(2x + y)$,
   Use $x = -1, y = 3$

7) $(2y)(2y - 3x)$,
   Use $x = 2, y = 1$

8) $(5x + 2y)y$,
   Use $x = 4, y = 2$

9) $3x + 2(2y - 2)$,
   Use $x = 4, y = 2$

10) $x + x(y - x)$,
    Use $x = 1, y = 2$

11) $2 + y(y - 2x)$,
    Use $x = 3, y = 1$

12) $x(2y - 3x)$,
    Use $x = -2, y = 2$

✎ Simplify each expression.

13) $2 + 2x - 3x =$
14) $4z + 2(z + 6) =$
15) $4y + 5 + 3y =$
16) $4w - 5 - 3w =$
17) $3m - 4(2m + 1) =$
18) $2t - 4(2 - t) =$
19) $-2k + 5 + 9k =$
20) $3(2d + 2) + (-6d) =$

21) $(-5)(5q + 3) - 3q =$
22) $(-a) + (-3)(1 - a) =$
23) $(2x - 1) - (6 - x) =$
24) $(-5)(2m + 2) + 3m =$
25) $9x - 4 - 5x + 3 =$
26) $(-3z) + (-2)(1 - z) =$
27) $14n - 3m + 12m - 19n =$
28) $33x - 5(6x - 1) =$

Simplifying Variable Expressions

✎ Simplify each expression.

1) $-6 - 2x^2 + 4x^2 =$

2) $3 + 10x^2 + 2 + 5x^2 =$

3) $4x^2 + 4(2x + 1) =$

4) $4x^2 - x(4x + 1) =$

5) $4(x^2 - 1) + 4x(2x + 1) =$

6) $(x^3 + 4) + x(2x^2 + 1) =$

7) $4x^2 - 19 + 4(2x^2 + 1) =$

8) $x^4 + 4(2x + 1) - 2x^4 =$

9) $(3x^2 + 1)x + (x^3 - x) =$

10) $(2x - 1)(2x + 1) =$

11) $(5x - 2)(5x + 2) =$

12) $(x + 4)(x + 4) =$

13) $2x^2 + 5x - 10x^2 - 2x =$

14) $32x - 15(2x + 2) =$

15) $(2x + 1)x + (-3x^2) - x =$

16) $2(2 - x) + 2(2x - 4) =$

17) $3(x + 9) =$

18) $(-6)(8x - 4) =$

19) $7x + 3 - 3x =$

20) $-2 - x^2 - 6x^2 =$

21) $3 + 10x^2 + 2 =$

22) $8x^2 + 6x + 7x^2 =$

23) $x^2 - 2x + 4x^2 - 1 =$

24) $x^3(2x - x^2 - 1) - x^5 =$

25) $Z^2 - Z(2Z + 5) =$

26) $4x^2 + 2x(3 - 5x) =$

27) $10m\left(\frac{m-15}{5}\right) + 4m =$

28) $3x^2 - 2x(4x + 1) =$

✎ Simplify.

29) $x(3x + 3), x = 2$

30) $4 - 5x + 9x - 3, x = 1$

31) $(2x + 3)(2x + 1), x = 2$

32) $x(3x - 14), x = 4$

33) $2x + 9 - 3x + 2, x = -2$

34) $x(x + 4) + 2x, x = 2$

35) $(15x - 25)x, x = 3$

36) $x + x(6x - 1), x = -1$

37) $x + (7 - x), x = 3$

38) $(3x - 1)(x + 2), x = -2$

Simplifying Polynomial Expressions

✍ Simplify each polynomial.

1) $(2x^3 + 5x^2) - (12x + 2x^2) =$

2) $(2x^5 + 2x^3) - (7x^3 + 6x^2) =$

3) $(12x^4 + 4x^2) - (2x^2 - 6x^4) =$

4) $14x - 3x^2 - 2(6x^2 + 6x^3) =$

5) $(5x^3 - 3) + 5(2x^2 - 3x^3) =$

6) $(4x^3 - 2x) - 2(4x^3 - 2x^4) =$

7) $2(4x - 3x^3) - 3(3x^3 + 4x^2) =$

8) $(2x^2 - 2x) - (2x^3 + 5x^2) =$

9) $2x^3 - (4x^4 + 2x) + x^2 =$

10) $x^4 - 2(x^2 + x) + 3x =$

11) $(2x^2 - x^4) - (4x^4 - x^2) =$

12) $4x^2 - 5x^3 + 15x^4 - 12x^3 =$

13) $2x^2 - 5x^4 + 14x^4 - 11x^3 =$

14) $2x^2 + 5x^3 - 7x^2 + 12x =$

15) $2x^4 - 5x^5 + 8x^4 - 8x^2 =$

16) $5x^3 + 15x - x^2 - 2x^3 =$

17) $14x^3 + 5 - 3(3x^2 + 1) =$

18) $2(3x + 1) + 2x(x - 2) =$

19) $-10x^2 + 4x^5 + 12x(1 + x) =$

20) $-3x - x^3 + x(3x + 3) =$

✍ Solve.

21) If $G = t^2 - 5t + 6$ and $H = -8t^2 + 7t - 9$ then what is the sum of sum G and H?

22) Subtract $6x^2 - 7x - 11$ from $5x^2 - 4x + 3$.

23) A polynomial of the 4th degree with a leading coefficient of 7 and a constant term of 8. Which answer is correct?

A: $7x^5 + 6x - 8$

B: $7x^4 + 6x + 8$

C: $7x^2 - 6x + 8$

D: $7x^4 + 8x + 7$

The Distributive Property

✍ *Use the distributive property to simply each expression.*

1) $x(3 - 2x) =$

2) $(-2)(2x - 1) + x =$

3) $3x(2 - x) =$

4) $(x + 1)(x - 1) =$

5) $(-5)(x + 3 - 3x) =$

6) $14(2x + 5) =$

7) $(x + 2)2x =$

8) $12(3x - 1) =$

9) $3x(x - 4) =$

10) $7x(1 + x) + 14 =$

11) $(-2x)x - 4x(4 + 5x) + 2 =$

12) $2x(4 - x) + (3x^2 + 4) =$

13) $(x + 1)(x - 1) + (-2x)x =$

14) $2x(x^2 + x + 1) + x(x - 1) =$

15) $3x(x - 1) - 3x(1 - x) =$

16) $(-2)(x - 1) + 10(x + 2) =$

17) $(3x + 1)(x - 1) + 2x^2 =$

18) $5(x + 1) + (-2x)(x + 2) =$

19) $x^2(x - 1) + x(2x^2 + 3) =$

20) $2(2 + 3x) =$

21) $3(5 + 5x) =$

22) $4(3x - 8) =$

23) $(6x - 2)(-2) =$

24) $(-3)(x + 2) =$

25) $(2 + 2x)5 =$

26) $(-4)(4 - 2x) =$

27) $-(-2 - 5x) =$

28) $(-6x + 2)(-1) =$

29) $(-5)(x - 2) =$

30) $-(7 - 3x) =$

31) $8(8 + 2x) =$

32) $2(12 + 2x) =$

33) $(-6x + 8)4 =$

34) $(3 - 6x)(-7) =$

35) $(-12)(2x + 1) =$

36) $(8 - 2x)9 =$

37) $5(7 + 9x) =$

38) $11(5x + 2) =$

39) $(-4x + 6)6 =$

40) $(3 - 6x)(-8) =$

41) $(-12)(2x - 3) =$

42) $(10 - 2x)9 =$

Evaluating One Variable

✎ **Simplify each algebraic expression.**

1) $4 - (2x + 1), x = 3$
2) $3x + 5, x = 2$
3) $-2x - 7, x = -2$
4) $3x - 5, x = -3$
5) $3x + 6, x = 3$
6) $12x - 10, x = 2$
7) $1 - 2x + 4, x = -3$
8) $5 - 3x, x = -3$
9) $\frac{15}{x+4} - 3, x = -7$
10) $\frac{x+2}{4} + 2x, x = 6$
11) $\frac{3x(x-2)}{(x+6)}, x = 3$
12) $\frac{4x+2}{x} + 4x, x = 2$
13) $\frac{8}{x+1} - 12x, x = 3$
14) $2x + \frac{x}{8}, x = 8$
15) $(2x + 1) + 3x, x = -2$
16) $2x + \frac{x+3}{2}, x = 3$
17) $(5x - 1) + (3x + 2), x = 3$
18) $(-x)(5x - 4) + 2x, x = 3$
19) $4x + 5 + 2x - 2, x = 2$
20) $(-2x)(x - 1), x = -2$

21) $-3x + 5(3 - x), x = 4$
22) $\frac{3x+5}{1-x} + 2x, x = 3$
23) $\frac{2(2x-1)}{(1-x)+2x+1}, x = -4$
24) $\frac{2x+5}{x} + 4x, x = 5$
25) $3x - \frac{1-x}{1+x}, x = -5$
26) $4x + 2 - 6x, x = 1$
27) $\frac{2x+5+14x}{x(1-x)}, x = 2$
28) $\frac{3x(1-2x)}{x+2}, x = 3$
29) $3x^2 + 2x - 1, x = -2$
30) $\frac{32x-20}{x+4}, x = -3$
31) $\frac{12x-10}{2x}, x = 1$
32) $(3x - 4) - 5x, x = -3$
33) $\frac{-2x+1-x}{x-3}, x = -3$
34) $4x(x + 1) + x, x = -4$
35) $2t(5 - t), t = -5$
36) $-2y(3y + 4), y = -2$
37) $4m + \frac{3m-1}{m+1}, m = 3$
38) $(3y + 1) + (2y - 3), y = 0$
39) $5G(3 + 2G) - G, G = -1$

Evaluating Two Variables

✍ *Simplify each algebraic expression.*

1) $2(x + 1) + y - 3 + 2$,
   $x = 3, y = 1$

2) $\left(-\frac{y+3}{x}\right) + 3y$
   $x = 5, y = 7$

3) $(-2a)(-2a - 2b)$,
   $a = -2, b = 3$

4) $2(x - 2y)$,
   $x = -2, y = 3$

5) $3x + 2 - 2y$,
   $x = 5, y = -2$

6) $2 + 2(-2x - 3y)$,
   $x = 4, y = 1$

7) $12(x + y + 1) + y$,
   $x = -1, y = 3$

8) $(2x + 1)y$
   $x = 3, y = 2$

9) $(x + 12) \div 2y$
   $x = 2, y = -1$

10) $(2x + y)2y + 2$,
    $x = 2, y = 5$

11) $2(x + y) + 5y$,
    $x = 2, y = 3$

12) $4y(2x + y)$,
    $x = -2, y = 2$

13) $2x + 5 - 3y$,
    $x = 2, y = -3$

14) $\frac{3x+5}{2y-2} + 2xy$,
    $x = -2, y = -1$

15) $3x + 2xy - 3y + 2$,
    $x = 4, y = 2$

16) $3y - \frac{42x}{2y} + 2y$,
    $x = 1, y = -3$

17) $2(2y - x) - 3(xy - 1)$,
    $x = 2, y = -3$

18) $\frac{y}{-x} + 2xy + 3y$,
    $x = -3, y = -6$

19) $2x - 3y + 2(x - y)$,
    $x = 4, y = -4$

20) $\frac{2x-y}{2x+y} - 2xy$,
    $x = -3, y = 2$

21) $4xy - 3x + y$,
    $x = 2, y = -1$

Combining like Terms

Simplify each expression.

1) $3x + 2 - 5x + 1 =$
2) $2x(1 + x) + 2 =$
3) $2(x - 1) + 3x - 1 =$
4) $2(2 - x) + 2x + 2 =$
5) $5x + 2 + 7x + 3x =$
6) $x + 2(3x - 2) =$
7) $(x + 1)(x - 1) - 2x =$
8) $9x - 2 - 5x + 7 =$
9) $2(2x + 1) - 5x =$
10) $2 + 2x - 5x - 3 =$
11) $12x - 2(1 - x) =$
12) $(3 - x)(x - 1) =$
13) $x + 1 + 3x + 4x =$
14) $3x + 9x - 2x + 7 =$
15) $-12x + 3 - x(-3) =$
16) $3x + 5x - 4x + 10 =$
17) $5(x - 2) + x(12 - 4) =$
18) $32x - 4 - 17x - 2x =$
19) $22x + 14x + 2 - 18x =$
20) $(x + 3) + 3x + 5x - 2 =$
21) $2 - 8x + 3 + 5x =$
22) $33x - 12x + 2x =$
23) $2(3x - 2) + (-4x + 4) =$
24) $2x - 4x + 7x + 2 =$
25) $72x - 33x + (-20x) =$
26) $3 - 5x - 12x + 25x =$
27) $12x - 5 + 4x - 3 =$
28) $12x + 4x - 21 =$
29) $5 + 2x - 8 =$
30) $(-2x + 6)2 =$
31) $7 + 3x + 6x - 4 =$
32) $9(x - 7x) - 5 =$
33) $7(3x + 6) + 2x =$
34) $3x - 12 - 5x =$
35) $2(4 + 3x) - 7x =$
36) $22x + 6 + 2x =$
37) $(-5x) + 12 + 7x =$
38) $(-3x) - 9 + 15x =$
39) $2(5x + 7) + 8x =$
40) $2(9 - 3x) - 17x =$
41) $-4x - (6 - 14x) =$
42) $(-4) - (3)(5x + 8) =$

Answers of Worksheets

Expressions and Variables

1) 20
2) 15
3) 5
4) 45
5) 0
6) −6
7) −8
8) 48
9) 16
10) 2
11) −3
12) −20
13) $-x + 2$
14) $6z + 12$
15) $7y + 5$
16) $w − 5$
17) $-5m − 4$
18) $6t − 8$
19) $7k + 5$
20) 6
21) $-28q − 15$
22) $2a − 3$
23) $3x − 7$
24) $-7m − 10$
25) $4x − 1$
26) $-z − 2$
27) $-5n + 9m$
28) $3x + 5$

Simplifying Variable Expressions

1) $2x^2 − 6$
2) $15x^2 + 5$
3) $4x^2 + 8x + 4$
4) $-x$
5) $12x^2 + 4x − 4$
6) $3x^2 + x + 4$
7) $12x^2 + x − 19$
8) $-x^4 + 8x + 4$
9) $4x^3$
10) $4x^2 − 1$
11) $25x^2 − 4$
12) $x^2 + 8x + 16$
13) $-8x^2 + 3x$
14) $-2x − 30$
15) $-x^2$
16) $2x − 4$
17) $3x + 27$
18) $-48x + 24$
19) $4x + 3$
20) $-7x^2 − 2$
21) $10x^2 + 5$
22) $15x^2 + 6x$
23) $5x^2 − 2x − 1$
24) $-2x^5 + 2x^4 − x^3$
25) $-Z^2 − 5Z$
26) $-6x^2 + 6x$
27) $2m^2 − 26m$
28) $-5x^2 − 2x$
29) 18
30) 5
31) 35
32) −8
33) 13
34) 16
35) 60
36) 6
37) 7
38) 0

Simplifying Polynomial Expressions

1) $2x^3 + 3x^2 − 12x$
2) $2x^5 − 5x^3 − 6x^2$
3) $18x^4 + 2x^2$
4) $-12x^3 − 15x^2 + 14x$
5) $-10x^3 + 10x^2 − 3$
6) $4x^4 − 4x^3 − 2x$
7) $-15x^3 − 12x^2 + 8x$
8) $-2x^3 − 3x^2 − 2x$
9) $-4x^4 + 2x^3 + x^2 − 2x$
10) $x^4 − 2x^2 + x$
11) $-5x^4 + 3x^2$
12) $15x^4 − 17x^3 + 4x^2$
13) $9x^4 − 11x^3 + 2x^2$
14) $5x^3 − 5x^2 + 12x$
15) $-5x^5 + 10x^4 − 8x^2$

16) $3x^3 - x^2 + 15x$
17) $14x^3 - 9x^2 + 2$
18) $2x^2 + 2x + 2$

19) $4x^5 + 2x^2 + 12x$
20) $-x^3 + 3x^2$
21) $-7t^2 + 2t - 3$

22) $x^2 - 3x - 14$
23) B

The Distributive Property

1) $-2x^2 - 3x$
2) $-3x + 2$
3) $-3x^2 + 6x$
4) $x^2 - 1$
5) $10x - 15$
6) $28x + 70$
7) $2x^2 + 4x$
8) $36x - 12$
9) $3x^2 - 12x$
10) $7x^2 + 7x + 14$
11) $-22x^2 - 16x + 2$
12) $x^2 + 8x + 4$
13) $2x^2 - x - 1$
14) $2x^3 + 3x^2 + x$
15) $6x^2 - 6x$

16) $8x + 22$
17) $5x^2 - 2x - 1$
18) $-2x^2 + x + 5$
19) $3x^3 - x^2 + 3x$
20) $6x + 4$
21) $15x + 15$
22) $12x - 32$
23) $-12x + 4$
24) $-3x - 6$
25) $10x + 10$
26) $8x - 16$
27) $5x + 2$
28) $6x - 2$
29) $-5x + 10$
30) $3x - 7$

31) $16x + 64$
32) $4x + 24$
33) $-24x + 32$
34) $42x - 21$
35) $-24x - 12$
36) $-18x + 72$
37) $45x + 35$
38) $55x + 22$
39) $-24x + 36$
40) $48x - 24$
41) $-24x + 36$
42) $-18x + 90$

Evaluating one Variables

1) -3
2) 11
3) -3
4) -14
5) 15
6) 14
7) 11
8) 14
9) -8
10) 14

11) 1
12) 13
13) -34
14) 17
15) -9
16) 9
17) 25
18) -27
19) 15
20) -12

21) -17
22) -1
23) 9
24) 23
25) $-13\frac{2}{4}$
26) 0
27) $-\frac{37}{2}$
28) -9
29) 7

30) -116
31) 1
32) 2
33) $-\frac{5}{3}$
34) 44
35) -100
36) -8
37) 14
38) -2
39) -4

Evaluating Two Variables

1) 8
2) 19
3) −8
4) −16
5) 21
6) −20
7) 39
8) 14
9) −7
10) 92
11) 25
12) −16
13) 18
14) $4\frac{1}{4}$
15) 24
16) −8
17) 5
18) 16
19) 36
20) 14
21) −15

Combining like Terms

1) $3 - 2x$
2) $2x^2 + 2x + 2$
3) $5x - 3$
4) 6
5) $15x + 2$
6) $7x - 4$
7) $x^2 - 4x + 1$
8) $4x + 5$
9) $2 - x$
10) $-3x - 1$
11) $14x - 2$
12) $-x^2 + 4x - 3$
13) $8x + 1$
14) $10x + 7$
15) $3 - 9x$
16) $4x + 10$
17) $13x - 10$
18) $13x - 4$
19) $18x + 2$
20) $9x + 1$
21) $5 - 3x$
22) $23x$
23) $2x$
24) $5x + 2$
25) $19x$
26) $8x + 3$
27) $16x - 8$
28) $16x - 21$
29) $2x - 3$
30) $-4x + 129x + 3$
31) $-54x - 5$
32) $23x + 42$
33) $-2x - 12$
34) $-x + 8$
35) $24x + 6$
36) $2x + 12$
37) $12x - 9$
38) $18x + 14$
39) $-23x + 18$
40) $10x - 6$
41) $-15x - 28$

Chapter 6:

Equations and Inequalities

Topics that you'll learn in this part:

- ✓ One–Step Equations
- ✓ One–Step Equation Word Problems
- ✓ Two–Step Equations
- ✓ Two–Step Equation Word Problems
- ✓ Multi–Step Equations
- ✓ Graphing Single–Variable Inequalities
- ✓ One–Step Inequalities
- ✓ Multi-Step Inequalities

One–Step Equations

✎Solve each equation.

1) $2x + 4 = 18$
2) $22 = (-8) + 3x$
3) $3x = (-30) + x$
4) $(-35) - x = (-6x)$
5) $(-6) = 4 + 10x$
6) $6 + 2x = (-2)$
7) $20x - 20 = (-220)$
8) $18 = 3x + 3$
9) $(-25) + 2x = (-17)$
10) $5x + 5 = (-45)$
11) $3x - 12 = (-21)$
12) $x - 2 = (-8)$
13) $(-30) = x - 18$
14) $8 = 2x - 2$
15) $(-6x) - 6 = 36$
16) $(-55) = (-5x) + 10$
17) $2x - 15 = 25$
18) $8x - 16 = 32$
19) $24 = (-6x)$
20) $8x + 4 = 68$
21) $50x + 50 = 300$

✎Write each sentence as an equation.

22) Eight less than $\frac{1}{3}$ a number M is -13.

23) A number of multiplied by -12.3 is -73.

24) Twice a number, decreased by twenty-nine, is seven.

25) Thirty-two is twice a number increased by eight.

26) Twelve is sixteen less than four times a number.

27) The sum of eight and a number is five less than seventy.

28) Ten less than a number z is twenty-five.

29) Seven less than a number is sixty.

30) The sum of ten and a number is two less than thirty.

31) Seventeen less than a number x is fifty-three.

32) The sum of L and 15 is eight less than eighty-two.

33) The sum of eight and a number is ten less than eighteen.

One–Step Equation Word Problems

Solve.

1) Thira has read 110 pages of a 290-page book. She reads 20 pages each day. How many days will it take to finish?

2) You and a friend split the cost of a moped rental. Your friend pays the bill. You owe your friend only $12, because your friend owed you $9 from yesterday. How much was the total bill?

3) Mr. Herman's class is selling candy for a school fundraiser. The class has a goal of raising $500 by selling c boxes of candy. For every box they sell, they make $5.5. How many boxes of candy they need to sell?

4) Lindsey is helping her uncle plant an apple orchard. After picking up a truckload of trees, they plant 20 of them. The next day, they still have 50 trees left to plant. Find the total number of trees in the original truckload?

5) Tina is baking chocolate chip cookies for a party at school. She leaves 12 at home for her family and brings the remaining 24 cookies to school to share with her classmates. Find the total number of cookies that Tina bakes?

6) The Laughing Lollipop candy store is holding a raffle, and Preston wins the grand prize! He wins a gift card to the store, as well as a bag of giant lollipops. He gives 21 lollipops to his friends, and he keeps the remaining 3 lollipops. Find the total number of lollipops in the bag?

Two–Step Equations

✍ Solve each equation.

1) $2(4 + x) = 20$
2) $(-2)(x + 3) = 42$
3) $3(2x - 4) = (-36)$
4) $6(2 - 2x) = 12$
5) $10(4x + 4) = (-60)$
6) $5(2x + 5) = 45$
7) $2(7 - 2x) = (-34)$
8) $(-4)(2x - 4) = 48$
9) $4(x - 5) = 8$
10) $\frac{2x - 10}{2} = 12$
11) $\frac{3x + 3}{9} = 10$
12) $15 = (-5)(3x - 6)$
13) $\frac{12 - x}{5} = 6$
14) $12 = (-12) + \frac{4x}{8}$
15) $\frac{24 + 3x}{4} = 9$
16) $(-2)(5 - 4x) = 70$
17) $(-11x) + 15 = 26$
18) $\frac{-24 - 8x}{6} = 8$
19) $\frac{2x - 12}{7} = 6$
20) $\frac{(-8) + 3x}{10} = \frac{2}{5}$
21) $\frac{2x + 2}{2} = 5$

✍ Fill in the blank with the appropriate number

22) $6 = \frac{x}{\Box} + 2, x = 16$
23) $\Box + \frac{x}{4} = -5, x = 4$
24) $0 = 4 + \frac{n}{\Box}, n = -20$
25) $-1 = \frac{\Box + x}{6}, x = -11$
26) $\frac{m + \Box}{3} = 8, m = 15$
27) $2(n + 5) = \Box, n = -6$
28) $144 = \Box(x + 5), x = -17$
29) $10 - 6x = \Box, x = 19$
30) $\frac{x + 5}{\Box} = -1, x = 11$
31) $-10 = \Box + 5x, x = 0$
32) $-10 = 10(x - \Box), x = 8$
33) $\frac{x}{9} + \Box = -2, x = -9$
34) $7(9 + x) = \Box, x = 3$
35) $8 + \frac{x}{\Box} = 5, x = 12$
36) $-243 = \Box(10 + x), x = 17$
37) $-15 = (\Box)x + 10, x = 5$
38) $-4 = \frac{x}{\Box} - 5, x = 20$

Two–Step Equation Word Problems

Solve.

1) Aliyah had $24 to spend on seven pencils. After buying them she had $10. How much did each pencil cost?

2) Maria bought seven boxes. A week later half of all her boxes were destroyed in a fire. There are now only 22 boxes left. With how many did she start?

3) Sara Wong spent half of his weekly allowance playing arcade games. To earn more money his parents let him weed the garden for $6.55. What is his weekly allowance if he ended with $11.01?

4) Rob had some paper with which to make note cards. On his way to his room he found two more pieces to use. In his room he cut each piece of paper in half. When he was done, he had 22 half-pieces of paper. With how many sheets of paper did he start?

5) The Cooking Club made some pies to sell during lunch to raise money for an end-of-year banquet. The cafeteria contributed two pies to the club. Each pie was then cut into seven pieces and sold. There was a total of 84 pieces to sell. How many pies did the club make?

6) Adam won 59 lollipops playing hoops at the county fair. At school he gave two to every student in his math class. He only has 3 remaining. How many students are in his class?

Multi–Step Equations

Solve each equation.

1) $2(2x - 1) = 10 + x$
2) $(-3)(2 - x) = 30 - 3x$
3) $2x + 14 = 3(10 + 2x)$
4) $3x - 24 = 2(10 - x) + x$
5) $10x + 4 = (-36) - x + x$
6) $12x + 1 = 8x - 59$
7) $15x - 12 - 12x = 7x$
8) $(-3x) + 2(x + 1) = 4(6 + x) + 3$

Solve.

9) Tickets to a fundraiser are $14 if purchased ahead of time and $25 if purchased at the door. The total amount raised from all ticket sales was $625. If eleven tickets were purchased at the door, how many tickets were purchased ahead of time?

10) On Friday, you raked leaves for 4 neighbors, on Saturday you raked leaves for 5 neighbors, and on Sunday you raked leaves for 3 neighbors. Over the three days you earned a total of $135. If you were paid the same amount at each house, write and solve an equation to determine how much you earned per house.

11) Your school is having a fundraiser. You are selling candy bars that have been donated by the Hershey Company. You set a personal goal of raising $200 for your school and met that goal. You sold a total of 120 candy bars and one neighbor gave you a $20 donation without taking any candy bars. Write and solve an equation to determine how much each candy bar sold for?

Graphing Single–Variable Inequalities

✎ *Draw a graph for each inequality.*

1) $x > 2$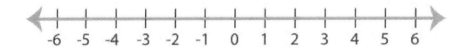

2) $x < 5$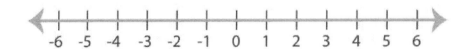

3) $x > -1$

4) $x > 3$

5) $x < -5$

6) $x > -2$

7) $x < 0$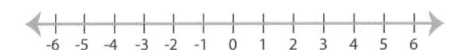

8) $x > 4$

One–Step Inequalities

✎ *Solve each inequality and graph it.*

1) $x + 2 \geq 3$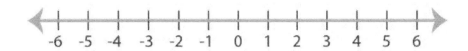

2) $x - 1 \leq 2$

3) $2x \geq 12$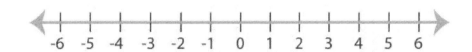

4) $4 + x \leq 5$

5) $x + 3 \leq -3$

6) $4x \geq 16$

7) $9x \leq 18$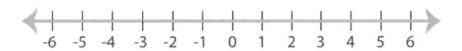

8) $x + 2 \geq 7$

Multi-Step Inequalities

✎ *Solve each inequality.*

1) $x - 2 \leq 6$

2) $3 - x \leq 3$

3) $2x - 4 \leq 8$

4) $3x - 5 \geq 16$

5) $x - 5 \geq 10$

6) $2x - 8 \leq 6$

7) $8x - 2 \leq 14$

8) $-5 + 3x \leq 10$

9) $2(x - 3) \leq 6$

10) $7x - 5 \leq 9$

11) $4x - 21 < 19$

12) $2x - 3 < 21$

13) $17 - 3x \geq -13$

14) $9 + 4x < 21$

15) $3 + 2x \geq 19$

16) $6 + 2x < 32$

17) $4x - 1 < 7$

18) $3(3 - 2x) \geq -15$

19) $-(3 + 4x) < 13$

20) $20 - 8x \geq -28$

21) $-3(x - 7) > 21$

22) $\dfrac{2x + 6}{4} \leq 10$

23) $\dfrac{4x + 8}{2} \leq 12$

24) $\dfrac{3x - 8}{7} > 1$

25) $4 + \dfrac{x}{3} < 7$

26) $\dfrac{9x}{7} - 7 < 2$

27) $\dfrac{4x + 12}{4} > 1$

28) $15 + \dfrac{x}{5} < 12$

Answers of Worksheets

One–Step Equations

1) 7
2) 10
3) −15
4) 7
5) −1
6) −4
7) −10
8) 5
9) 4
10) −10
11) −3
12) −6
13) −12
14) 5
15) −5
16) 13
17) 20
18) 6
19) −4
20) 8
21) 5
22) $8 - \frac{1}{3}M = -13$
23) $(-12.3)f = -73$
24) $2x - 29 = 7$
25) $2x + 8 = 32$
26) $16 - 4x = 12$
27) $8 + x = 5 - 70$
28) $10 - z = 25$
29) $7 - x = 60$
30) $10 + x = 2 - 30$
31) $17 - x = 53$
32) $l + 15 = 8 - 82$
33) $8 + x = 10 - 8$

One–Step Equation Word Problems

1) 9
2) 42
3) 91
4) 30
5) 36
6) 24

Two–Step Equations

1) 6
2) −24
3) −4
4) 0
5) −2.5
6) 2
7) 12
8) −4
9) 7
10) 17
11) 29
12) 1
13) −18
14) 48
15) 4
16) 10
17) −1
18) −9
19) 27
20) 4
21) 4
22) 4
23) −6
24) 5
25) 5
26) 9
27) −2
28) −12
29) −104
30) −16
31) −10
32) −9
33) −1
34) 84
35) −4
36) −9
37) −1
38) 20

Two–Step Equation Word Problems

1) $2
2) 37
3) $8.92
4) 9
5) 10
6) 28

Multi–Step Equations

1) 4
2) 6
3) −4
4) 11
5) −4
6) −15
7) −3
8) −5
9) 25
10) 11.25
11) 1.5

Graphing Single–Variable Inequalities

1)
2)
3)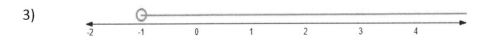
4)
5)
6)
7)
8)

One–Step Inequalities

1)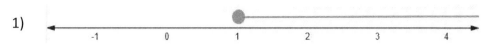

2)

3)

4)

5)

6)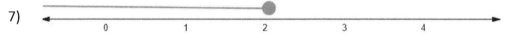

7)

8)

Multi-Step Inequalities

1) $x \leq 8$
2) $x \geq 0$
3) $x \leq 6$
4) $x \geq 7$
5) $x \geq 15$
6) $x \leq 7$
7) $x \leq 2$
8) $x \leq 5$
9) $x \leq 6$
10) $x \leq 2$
11) $x < 10$
12) $x < 12$
13) $x \leq 10$
14) $x < 3$
15) $x \geq 8$
16) $x < 13$
17) $x < 2$
18) $x \leq 4$
19) $x > -4$
20) $x \leq 6$
21) $x < 0$
22) $x \leq 17$
23) $x \leq 4$
24) $x > 5$
25) $x < 9$
26) $x < 7$
27) $x > -2$
28) $x < -15$

Chapter 7:

Systems of Equations

Topics that you'll learn in this part:

- ✓ Solving Systems of Equations by Substitution
- ✓ Solving Systems of Equations by Elimination
- ✓ Systems of Equations Word Problems

Solving Systems of Equations by Substitution

✍ *Solve each system of equation by substitution.*

1) $\begin{cases} x + 4 = y \\ 2x + y = 1 \end{cases}$

2) $\begin{cases} 2x + y = 2 \\ x - y = 10 \end{cases}$

3) $\begin{cases} 4y - 2 = x \\ x + 2y = 4 \end{cases}$

4) $\begin{cases} 1 - x = 2y \\ 2x - 2y = -16 \end{cases}$

5) $\begin{cases} 3x - y = 6 \\ 2x + y = 14 \end{cases}$

6) $\begin{cases} 2x + 1 = 1 - y \\ x - 3y = 28 \end{cases}$

7) $\begin{cases} y - 3x = 4 \\ 2(x + y) + 5 = -11 \end{cases}$

8) $\begin{cases} 9 - 6x = 3y \\ 3x + y = 1 \end{cases}$

9) $\begin{cases} y = 7x - 10 \\ y = -3 \end{cases}$

10) $\begin{cases} y = -8x \\ 2x + 4y = 0 \end{cases}$

11) $\begin{cases} 6x - 11 = y \\ -2x - 3y = -7 \end{cases}$

12) $\begin{cases} 2x - 3y = -1 \\ x - 1 = y \end{cases}$

13) $\begin{cases} -3x - 3y = 3 \\ -5x - 17 = y \end{cases}$

14) $\begin{cases} y = -3x + 5 \\ 5x - 4y = -3 \end{cases}$

15) $\begin{cases} y - 5x = -7 \\ 12 - 2y = 3x \end{cases}$

16) $\begin{cases} y = 6 + 4x \\ -5x - y = 21 \end{cases}$

17) $\begin{cases} -7x - 2y = -13 \\ x - 2y = 11 \end{cases}$

18) $\begin{cases} -5x + y = -2 \\ 12 + 6y = 3x \end{cases}$

19) $\begin{cases} 3 + y - 5x = 0 \\ 3x - 8y = 24 \end{cases}$

20) $\begin{cases} x + 3y = 1 \\ -3x - 3y = -15 \end{cases}$

21) $\begin{cases} -3x - 8y = 20 \\ -5x + y = 19 \end{cases}$

22) $\begin{cases} 3y - 3x = 3 \\ y - 5x = 13 \end{cases}$

23) $\begin{cases} 6x + 6y = -6 \\ 5x + y = -13 \end{cases}$

24) $\begin{cases} 2x + y = 20 \\ 6x - 5y = 12 \end{cases}$

25) $\begin{cases} -3x - 4y = 2 \\ 3x + 3y = -3 \end{cases}$

26) $\begin{cases} -2x + 6y = 6 \\ -7x + 8y = -5 \end{cases}$

27) $\begin{cases} 4x + y = 24 \\ y = 4x + 24 \end{cases}$

28) $\begin{cases} -5x + 2y = 9 \\ y = 7x \end{cases}$

29) $\begin{cases} -7x - 6y - 4 = 0 \\ x - 8 = -3y \end{cases}$

30) $\begin{cases} -4x = 20 - 7y \\ y = 3x + 15 \end{cases}$

Solving Systems of Equations by Elimination

✍ *Solve each system of equation by elimination.*

1) $\begin{cases} -4x - 2y = -12 \\ 4x + 8y = -24 \end{cases}$

2) $\begin{cases} 4x + 8y = 20 \\ -4x + 2y = -30 \end{cases}$

3) $\begin{cases} x - y = 11 \\ 2x + y = 19 \end{cases}$

4) $\begin{cases} -6x + 5y = 1 \\ 6x + 4y = -10 \end{cases}$

5) $\begin{cases} -2x + 9y = -25 \\ -4x - 9y = -23 \end{cases}$

6) $\begin{cases} 8x + y = -16 \\ -3x + y = -5 \end{cases}$

7) $\begin{cases} -6x + 6y = 6 \\ -6x + 3y = -12 \end{cases}$

8) $\begin{cases} 7x + 2y = 24 \\ 8x + 2y = 30 \end{cases}$

9) $\begin{cases} 5x + y = 9 \\ 10x - 7y = -18 \end{cases}$

10) $\begin{cases} -4x + 9y = 9 \\ x - 3y = -6 \end{cases}$

11) $\begin{cases} -3x + 7y = -16 \\ -9x + 5y = 16 \end{cases}$

12) $\begin{cases} -7x + y = -19 \\ -2x + 3y = -19 \end{cases}$

13) $\begin{cases} 16x - 10y = 10 \\ -8x - 6y = 6 \end{cases}$

14) $\begin{cases} 8x + 14y = 4 \\ -6x - 7y = -10 \end{cases}$

15) $\begin{cases} -4x - 15y = -17 \\ -x + 5y = -13 \end{cases}$

16) $\begin{cases} -x - 7y = 14 \\ -4x - 14y = 28 \end{cases}$

17) $\begin{cases} -7x - 8y = 9 \\ -4x + 9y = -22 \end{cases}$

18) $\begin{cases} 5x + 4y = -30 \\ 3x - 9y = -18 \end{cases}$

19) $\begin{cases} -4x - 2y = 14 \\ -10x + 7y = -25 \end{cases}$

20) $\begin{cases} 3x - 2y = 2 \\ 5x - 5y = 10 \end{cases}$

21) $\begin{cases} 5x + 4y = -14 \\ 3x + 6y = 6 \end{cases}$

22) $\begin{cases} 2x + 10y = 6 \\ -5x - 20y = -15 \end{cases}$

23) $\begin{cases} -7x - 20y = -14 \\ 10y + 4 = 2x \end{cases}$

24) $\begin{cases} 3 + 2x = y \\ -3 - 7y = 10x \end{cases}$

25) $\begin{cases} -10x + 3y = 5 \\ x - y = -4 \end{cases}$

26) $\begin{cases} 12x - 5y = -20 \\ y = x + 4 \end{cases}$

27) $\begin{cases} -4x + 11y = 15 \\ x - 2y = 0 \end{cases}$

28) $\begin{cases} 7x - 3y = 20 \\ y = 5x - 4 \end{cases}$

29) $\begin{cases} -5x + 4y = 3 \\ x + 15 = 2y \end{cases}$

30) $\begin{cases} 8x + 5y = 24 \\ y = -4x \end{cases}$

Systems of Equations Word Problems

Solve.

1) A used book store also started selling used CDs and videos. In the first week, the store sold a combination of 40 CDs and videos. They charged $4 per CD and $6 per video and the total sales were $180. Determine the total number of CDs and videos sold.

2) At the end of the 2000-2001 football season, 31 Super Bowl games had been played with the current two football leagues, the American Football Conference (AFC) and the National Football Conference (NFC). The NFC won five more games than the AFC. Determine the total number of wins by each conference.

3) The length of Sally's garden is 4 meters greater than 3 times the width. The perimeter of her garden is 72 meters. Find the dimensions of Sally's garden.

4) Giselle works as a carpenter and as a blacksmith. She earns $20 as a carpenter and $25 as a blacksmith. Last week, Giselle worked both jobs for a total of 30 hours and earned a total of $690. How long did Giselle work as a carpenter last week, and how long did she work as a blacksmith?

5) At a sale on winter clothing, Cody bought two pairs of gloves and four hats for $43.00. Tori bought two pairs of gloves and two hats for $30.00. Find the prices of the hats and gloves.

Answers of Worksheets

Solving Systems of Equations by Substitution

1) $(-1, 3)$
2) $(4, -6)$
3) $(2, 1)$
4) $(-5, 3)$
5) $(4, 6)$
6) $(4, -8)$
7) $(-3, -5)$
8) $(-2, 7)$
9) $(1, -3)$
10) $(0, 0)$
11) $(2, 1)$
12) $(4, 3)$
13) $(-4, 3)$
14) $(1, 2)$
15) $(2, 3)$
16) $(-3, -6)$
17) $(3, -4)$
18) $(0, -2)$
19) $(0, -3)$
20) $(7, -2)$
21) $(-4, -1)$
22) $(-3, -2)$
23) $(-3, 2)$
24) $(7, 6)$
25) $(-2, 1)$
26) $(3, 2)$
27) $(0, 24)$
28) $(1, 7)$
29) $(-4, 4)$
30) $(-5, 0)$

Solving Systems of Equations by Elimination

1) $(6, -6)$
2) $(7, -1)$
3) $(10, -1)$
4) $(-1, -1)$
5) $(8, -1)$
6) $(-1, -8)$
7) $(5, 6)$
8) $(6, -9)$
9) $(1, 4)$
10) $(9, 5)$
11) $(-4, -4)$
12) $(2, -5)$
13) $(0, -1)$
14) $(4, -2)$
15) $(8, -1)$
16) $(0, -2)$
17) $(1, -2)$
18) $(-6, 0)$
19) $(-1, -5)$
20) $(-2, -4)$
21) $(-6, 4)$
22) $(3, 0)$
23) $(2, 0)$
24) $(-1, 1)$
25) $(1, 5)$
26) $(0, 4)$
27) $(10, 5)$
28) $(-1, -9)$
29) $(9, 12)$
30) $(-2, 8)$

Systems of Equations Word Problems

1) $(10\ videos, 30 CDs)$
2) $(NFC\ wins\ 18, AFC\ wins\ 13)$
3) $(8, 28)$
4) $(12, 18)$
5) $(gloves: \$6, hats: \$8.5)$

Chapter 8:

Linear Functions

Topics that you'll learn in this part:

- ✓ Finding Slope
- ✓ Graphing Lines Using Slope–Intercept Form
- ✓ Graphing Lines Using Standard Form
- ✓ Writing Linear Equations
- ✓ Graphing Linear Inequalities
- ✓ Finding Midpoint
- ✓ Finding Distance of Two Points
- ✓ Slope and Rate of Change
- ✓ Find the Slope, X–intercept and Y–intercept
- ✓ Write an Equation from a Graph
- ✓ Slope–intercept form
- ✓ Point–slope form
- ✓ Equations of horizontal and vertical lines
- ✓ Equation of parallel or perpendicular lines

Finding Slope

✍ Find the slope of the line through each pair of points.

1) $(0, 2), (-3, 2)$

2) $(2, 2), (-3, -2)$

3) $(0, 0), (4, -2)$

4) $(2, 5), (1, 1)$

5) $(15, 8), (15, -8)$

6) $(-2, 0), (1, 3)$

7) $(-2, -4), (2, 3)$

8) $(-2, -4), (-1, 0)$

9) $(-6, 5), (-1, 0)$

10) $(-6, 10), (4, 0)$

11) $(15, 5), (-5, 1)$

12) $(-10, -2), (0, 3)$

13) $(5, -2), (3, 3)$

14) $(5, 0), (10, 10)$

15) $(0, 2), (2, 10)$

16) $(0, 0), (-10, 10)$

17) $(-22, -12), (-5, 5)$

18) $(0, 0), (-10, 30)$

19) $(12, 12), (-3, 6)$

20) $(-6, 2), (-4, 6)$

21) $(-2, 7), (-8, 7)$

22) $(14, 70), (28, 7)$

23) $(9, 6), (-1, -1)$

24) $(0, 6), (12, 24)$

✍ Solve.

25) Some building codes require the slope of a stairway to be no steeper than 0.88 or $\frac{22}{25}$. The stairs in the Adam's house measure 11-inch-deep and 6 inch high. Do the stairs meet the code requirement?

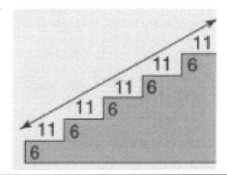

Graphing Lines Using Slope–Intercept Form

✏️ **Sketch the graph of each line.**

1) $y = \frac{1}{2}x - 4$

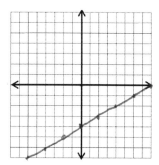

2) $y = x + 1$

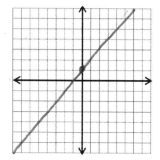

3) $y = -x + 1$

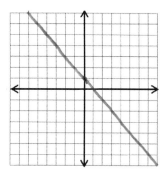

4) $y = 3x - 1$

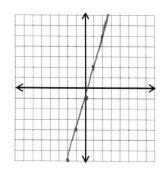

5) $y = -\frac{1}{5}x + \frac{1}{3}$

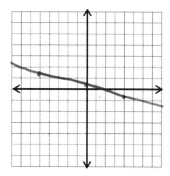

6) $y = \frac{3}{5}x - 1$

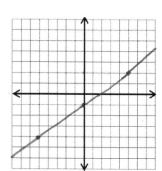

Graphing Lines Using Standard Form

✎ Sketch the graph of each line.

1) $2x - 2y = 4$

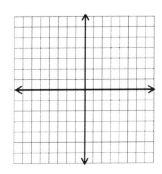

2) $3x + y = 2$

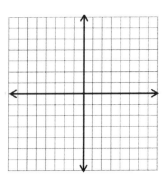

3) $x - y = 10$

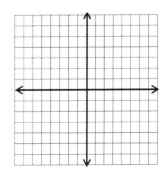

4) $3x + 2y = 2$

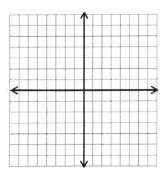

5) $y - 2x = 6$

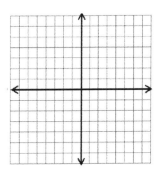

6) $2x - 3y = 6$

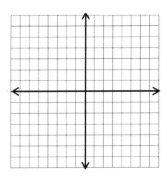

Writing Linear Equations

✎ **Write the slope–intercept form of the equation of the line through the given points.**

1) through: $(0, 2), (2, 4)$
2) through: $(0, 1), (2, -3)$
3) through: $(0, 2), (-2, -4)$
4) through: $(0, 2), (2, -8)$
5) through: $(0, 4), (-8, 0)$
6) through: $(1, 0), (0, -3)$
7) through: $(0, -2), (3, 0)$
8) through: $(0, 1), (-1, -3)$
9) through: $(1, -1), (1, 5)$
10) through: $(0, -0.5), (2, 3.5)$

11) through: $(0, 5), (-3, -4)$
12) through: $(5, 0), (0, 5)$
13) through: $(1, -6), (1.5, 0)$
14) through: $(0, 2), (1, -7)$
15) through: $(0, 3), (3, -3)$
16) through: $(0, 6), (5, -4)$
17) through: $(0, 3), (2, -5)$
18) through: $(4, 3), (2, 6)$
19) through: $(0, 1), (-4, 11)$
20) through: $(1.5, 2), (4, 5.5)$

✎ **Write the line slope from graph.**

21)

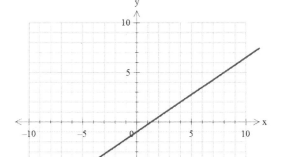

22)

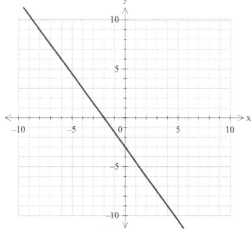

Graphing Linear Inequalities

✎ **Sketch the graph of each linear inequality.**

1) $y > 3x + 4$

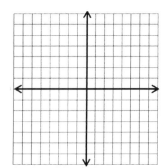

2) $y < -2x - 1$

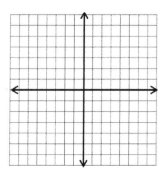

3) $y + \frac{1}{2} \geq \frac{1}{2}x$

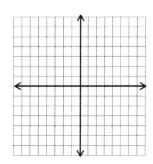

4) $y \leq 3 - 2x$

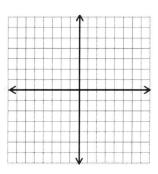

5) $4x - 2 \leq y$

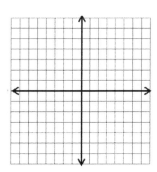

6) $1 - y \geq x$

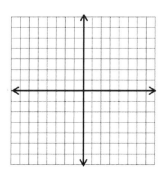

Finding Midpoint

✏ Find the midpoint of the line segment with the given endpoints.

1) $(-4, 5), (3, -\frac{1}{2})$
2) $(3, 7), (5, -3)$
3) $(-4, -2), (1, -10)$
4) $(3, -\frac{3}{2}), (4, 2)$
5) $(7, 0), (0, -10)$
6) $(4, -9), (0, 0)$
7) $(-3, -10), (3, 3)$
8) $(9, 1), (4, 4)$

9) $(75, 80), (40, 0)$
10) $(0, 13), (13, 0)$
11) $(-10, -3), (15, 14)$
12) $(33, 13), (9, 11)$
13) $(15, 0), (-10, -1)$
14) $(-3, 3), (15, 20)$
15) $(1, 10), (33, 0.5)$
16) $(90, -90), (50, 0)$

✏ Solve.

17) Find the midpoint of $\overline{AB}$ using the information in the diagram.

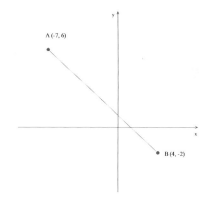

18) See the diagram. Find the midpoint of $\overline{BC}$.

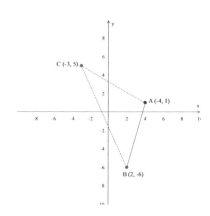

Finding Distance of Two Points

✎ *Find the distance between each pair of points.*

1) $(5, 7), (5, 3)$
2) $(6, 0), (-4, -10)$
3) $(-2, 1), (10, -5)$
4) $(33, -5), (17, 8)$
5) $(-6, -5), (6, 5)$

6) $(0, 0), (5, 8)$
7) $(3, 4), (0, 0)$
8) $(12, 16) (0, 0)$
9) $(-17, 1), (2, -6)$
10) $(-3, 0), (14, 0)$

✎ *Solve.*

11) Camp Sunshine is also on the lake. Use the Pythagorean Theorem to find the distance between Gabriela's house and Camp Sunshine to the nearest tenth of a meter.

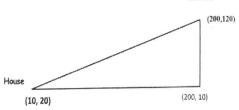

12) The class of math is mapped on a coordinate grid with the origin being at the center point of the hall. Mary's seat is located at the point (-4,7) and Betty's seat is located at (-2, 5). How far is it from Mary's seat to Betty's seat?

13) The teaching building of a university is mapped on a coordinate grid with the origin being at library. Math's building is located at the point (1,5) and History's building is located at (4, 9). How far is it from Math's building to History's building?

Slope and Rate of Change

✏️ **Find the slope of the line that passes through the points.**

1) $(-10, 4), (0, 2)$
2) $(3, 1), (12, 0)$
3) $(15, 0), (2, 13)$
4) $(12, 97), (-3, -5)$
5) $(-10, -8), (3, 1)$
6) $(-17, 20), (-1, -1)$
7) $(-2, -2), (0, 8)$
8) $(15, -8), (-1, 8)$
9) $(13, 0), (-3, 11)$
10) $(4, 3), (5, 1)$
11) $(12, 3), (-4, 3)$
12) $(1, -1), (0, 0)$

✏️ **Find the value of r so the line that passes through each pair of points has the given slope.**

13) $(1, 1), (2, r), m = 2$
14) $(-1, r), (0, 3), m = 1$
15) $(3, -1), (r, 3), m = -4$
16) $(r, -1), (0, 5), m = 3$
17) $(5, 1), (2, r), m = -1$
18) $(-3, 1), (r, 4), m = 3$
19) $(6, 2), (r, 4), m = 2$
20) $(6, r), (3, 4), m = -3$
21) $(12, -9), (r, -8), m = -1$
22) $(7, r), (5, -2), m = 3$
23) $(1, 1), (r, 5), m = 2$
24) $(7, r), (5, 10), m = -11$
25) $(1, 1), (r, 5), m = 2$
26) $(4, r), (-3, -8), m = \frac{2}{7}$
27) $(7, -12), (5, r), m = -11$
28) $(19, 3), (20, r), m = 0$
29) $(15, 8), (r, 9), m = -\frac{1}{32}$
30) $(r, -12), (15, -3), m = 1$
31) $(3, 1), (r, -5), m = -\frac{3}{2}$
32) $(3, -2), (-7, r), m = -1$
33) $(20, r), (-20, -10), m = 0.875$
34) $(6, 0), (r, -2), m = 0.2$
35) $(200, 100), (100, r), m = -1$
36) $(0, r), (10, 0), m = -0.5$
37) $(2, -8), (-10, r), m = -0.5$

Find the Slope, x–intercept and y–intercept

✏️ **Find the x and y intercepts for the following equations.**

1) $x + 2y = -2$

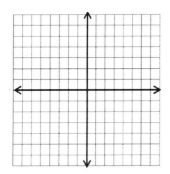

2) $y = 2x - 3$

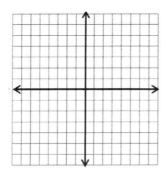

3) $-3x = 3y + 1$

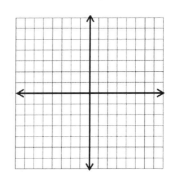

4) $2 - 2y = -5x$

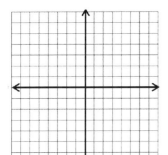

5) $-5x + 2y = 10$

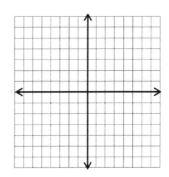

6) $2y = 2 - x$

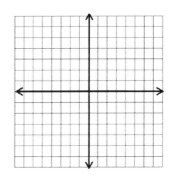

7) $1 - 2y = 7x$

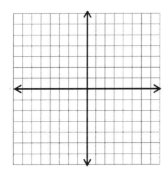

8) $2x + 4y = 9$

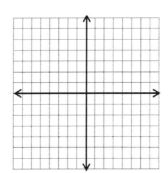

9) $3y = 7 + 2x$

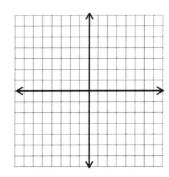

Write an equation from a graph

✎ **Write the slope intercept form of the equation of each line.**

1)

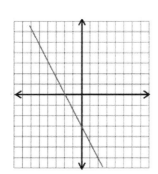

2)

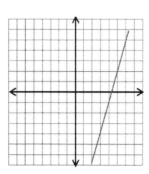

3)

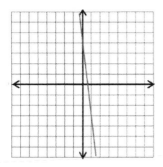

4)

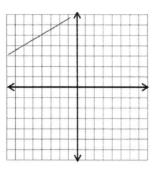

5)

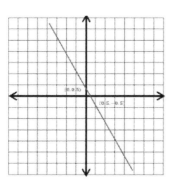

6)

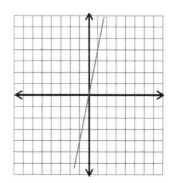

Slope–intercept Form

✏️ *Write the slope–intercept form of the equation of each line.*

1) $3x - 12 = 12y$

2) $2x - 3y = 6$

3) $\frac{-3}{2}x + 5y = 5$

4) $3(2x + y) = 6$

5) $14x + 7y = 28$

6) $3x - 4y = 5y + 1$

7) $3x + 4y = 3 + y$

8) $4 - 2x + y = 2$

9) $3x - y = 2$

10) $4y - 5x = -5$

11) $7x + 4y = 14 - 3y$

12) $15y - 5x = 30$

✏️ *Solve.*

13) Suppose that the water level of a river is 34 feet and that it is receding at a rate of 0.5 foot per day. Find the slope and write a sentence to interpret the slope in detail. Write an equation for the water level, L, after d days. In how many days will the water level be 26 feet?

14) For babysitting, Nicole charges a flat fee of $3, plus $5 per hour. Write an equation for the cost, C, after h hours of babysitting. What do you think the slope and the y-intercept represent? How much money will she make if she baby-sits 5 hours?

15) In order to "curve" a set of test scores, a teacher uses the equation $y = 2.5x + 10$, where y is the curved test score and x is the number of problems answered correctly. Find the test score of a student who answers 32 problems correctly. Explain what the slope and the y-intercept mean in the equation.

Point–slope Form

✎ **Find the slope of the following lines. Name a point on each line.**

1) $y = 3(x + 3)$

2) $y = 3x - 2$

3) $y = 4x + 3$

4) $y = \frac{2}{3}x - 1$

5) $y + 3 = 4x + 8$

6) $2y = 3x - 4$

7) $y - 1 = 4x - 2$

8) $3y - 4 = 2x$

9) $y + 1 = 3x - 3$

10) $4y - 8 = 4x + 10$

11) $y + 1 = 32(x + 1)$

12) $15x + 3 = 3y$

✎ **Write an equation in point–slope form for the line that passes through the given point with the slope provided.**

13) $(3,2), m = \frac{1}{2}$

14) $(1,-2), m = 3$

15) $(3,-2), m = -2$

16) $(4,1), m = 3$

17) $(-2,-3), m = -\frac{1}{2}$

18) $\left(\frac{3}{4}, -\frac{1}{2}\right), m = 3$

19) $(-2,0) \, m = \frac{3}{4}$

20) $\left(-\frac{3}{7}, \frac{1}{2}\right), m = 2$

21) $(0,2), m = 5$

22) $(-3,1), m = \frac{4}{5}$

23) $\left(-\frac{2}{3}, \frac{2}{3}\right), m = 4$

24) $(5,0), m = \frac{1}{5}$

25) $(3,3), m = 10$

26) $(-6,-5), m = -1$

27) $(-6,0), m = -\frac{2}{3}$

28) $(1,1), m = -4$

29) $(-3,2), m = -4$

30) $\left(-\frac{1}{2}, \frac{1}{4}\right), m = \frac{1}{2}$

31) $(-4,2), m = \frac{7}{3}$

32) $(0,1), m = \frac{1}{2}$

33) $(4,1), m = 2$

34) $(3,3), m = -1$

35) $(-1,0), m = -2$

36) $(6,5), m = 4$

37) $(1,-2), m = 3$

Equations of Horizontal and Vertical Lines

✏️ *Sketch the graph of each line.*

1) $y = 0$

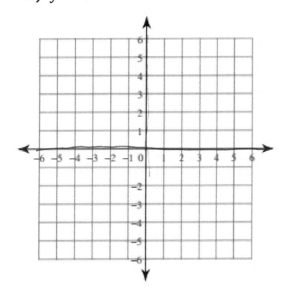

2) $y = 2$

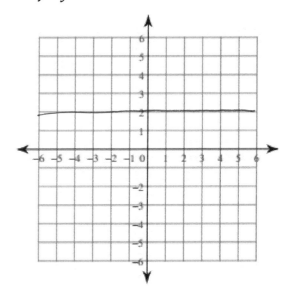

3) $x = -4$

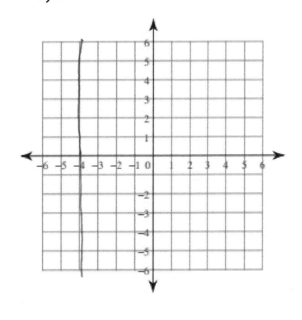

4) $x = 3$

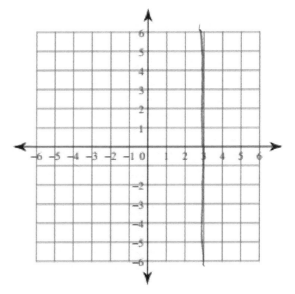

Equation of Parallel or Perpendicular Lines

✍ *Write an equation of the line that passes through the given point and is parallel to the given line.*

1) $(-2,3), 2y + 1 = 3x$
2) $(0,7), 3y + x = 2$
3) $(6,-1), 3y - x = 1$
4) $(-2,4), y + 2x = 2$
5) $(3,2), 2y + x = -2$

6) $(1,6) - 2y = x + 1$
7) $(3,2), y = 1 - x$
8) $(2,2), y = 3x$
9) $(-3,0), y = 2x - 2$
10) $(-3,2), 3y - 2 = 3x$

✍ *Write an equation of the line that passes through the given point and is perpendicular to the given line.*

11) $(-2,1), y + 1 = 3x$
12) $(0,-7), y + 2x = 2$
13) $(0,-1), 2y - 2x = 1$
14) $(-3,3), 2y + 2x = 2$

15) $(5,0), 3x + y = 5$
16) $(1,2), y = 3x + 1$
17) $(-3,0), 2y + 3 = x$
18) $(4,3), y = -5x + 1$

✍ *Solve.*

19) A caterer charges $120 to cater a party for 15 people and $200 for 25 people. Assume that the cost, y, is a linear function of the number of x people. Write an equation in slope-intercept form for this function. What does the slope represent? How much would a party for 40 people cost?

   A. $280
   B. $330

   C. $300
   D. $320

Answers of Worksheets

Finding Slope

1) 0
2) 0.8
3) −0.5
4) 4
5) Undefined

6) 1
7) 1.75
8) 4
9) −1
10) −1

11) 0.2
12) 0.5
13) −2.5
14) 2
15) 4

16) −1
17) 1
18) −3
19) 0.4
20) 2

21) 0
22) −4.5
23) 0.7
24) 1.5
25) 2.5

Graphing Lines Using Slope–Intercept Form

1)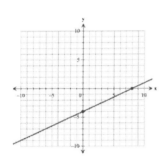

2)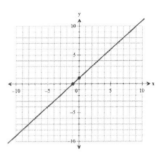

3)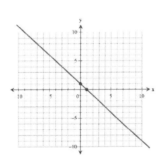

4)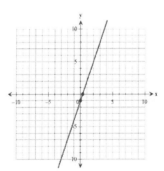

5)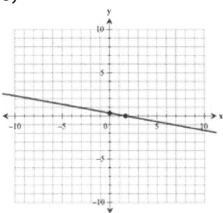

6)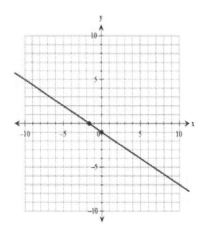

Graphing Lines Using Standard Form

1)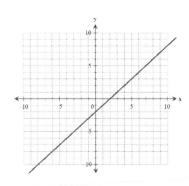

2)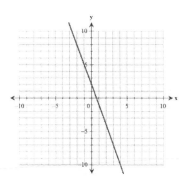

3)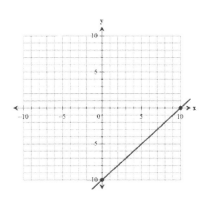

4)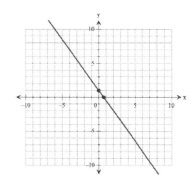

5)

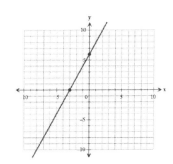

6)

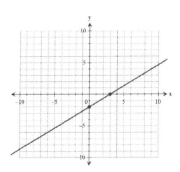

Writing Linear Equations

1) $y = x + 2$
2) $y = 1 - 2x$
3) $y = 3x + 2$
4) $y = 2 - 5x$
5) $y = \frac{1}{2}x + 4$
6) $y = 3x - 3$
7) $y = \frac{2}{3}x - 2$
8) $y = 4x + 1$

9) $x = 1$
10) $y = 2x - \frac{1}{2}$
11) $y = 3x + 5$
12) $y = -x + 5$
13) $y = 12x - 18$
14) $y = -9x + 2$
15) $y = 3 - 2x$
16) $y = 6 - 2x$

17) $y = 3 - 4x$
18) $y = 9 - \frac{3}{2}x$
19) $y = 1 - 2\frac{1}{2}x$
20) $y = 1.4x - 0.1$
21) $y = 0.75x - 1$
22) $y = -1.5x - 3$

Graphing Linear Inequalities

1)

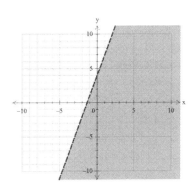

2)

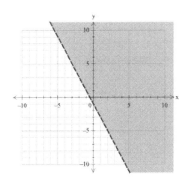

3)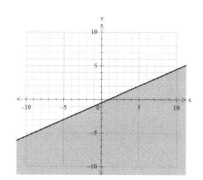

4)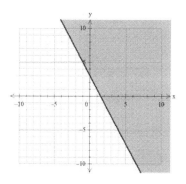

5)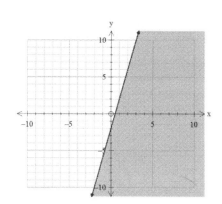

6)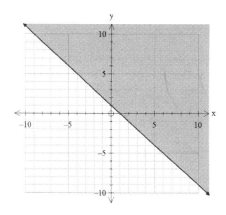

Finding Midpoint

1) $(-0.5, 2.25)$
2) $(4, 2)$
3) $(-1.5, -6)$
4) $(3.5, 0.25)$
5) $(3.5, -5)$

6) $(2, -4.5)$
7) $(0, -3.5)$
8) $(6.5, 2.5)$
9) $(57.5, 40)$
10) $(6.5, 6.5)$

11) $(2.5, 5.5)$
12) $(21, 12)$
13) $(2.5, -0.5)$
14) $(6, 11.5)$
15) $(17, 5.25)$

16) $(70, -45)$
17) $(-1.5, 2)$
18) $(-0.5, -0.5)$

Finding Distance of Two Points

1) 4
2) 14.14
3) 13.41

4) 20.62
5) 15.62
6) 9.43

7) 5
8) 20
9) 20.25

10) 17
11) 214.7
12) $2\sqrt{2}$

13) 5

Slope and Rate of Change

1) -0.2
2) $-\frac{1}{9}$
3) -1
4) $\frac{34}{5}$
5) $\frac{9}{13}$
6) $-\frac{7}{2}$
7) 5
8) -1
9) $-\frac{11}{16}$
10) -2
11) 0
12) -1
13) 3
14) 2
15) 2
16) -2
17) 4
18) -2
19) 7
20) -5
21) 11
22) 4
23) 3
24) -12
25) 3
26) -6
27) 10
28) 3
29) -17
30) 6
31) 7
32) 8
33) 25
34) -4
35) 200
36) 5
37) -2

Find the Slope, x–intercept and y–intercept

1) $y = -1, x = -2$
2) $y = -3, x = \frac{3}{2}$
3) $y = -\frac{1}{3}, x = -\frac{1}{3}$
4) $y = 1, x = -\frac{2}{5}$
5) $y = 5, x = -2$
6) $y = 1, x = 2$
7) $y = \frac{1}{2}, x = 2$
8) $y = \frac{9}{4}, x = \frac{9}{2}$
9) $y = \frac{7}{3}, x = -\frac{7}{2}$

Write an equation from a graph

1) $y = -\frac{3}{2}x - 3$
2) $y = 3x - 13$
3) $y = -7x + 4$
4) $y = \frac{1}{2}x + 7$
5) $y = -2x + \frac{1}{2}$
6) $y = 4x$

Slope–intercept form

1) $y = \frac{1}{4}x - 4$
2) $y = \frac{2}{3}x - 2$
3) $y = \frac{3}{10}x + 1$
4) $y = -2x + 2$
5) $y = -2x + 4$
6) $y = \frac{1}{3}x - \frac{1}{9}$
7) $y = -x + 1$
8) $y = 2x - 2$
9) $y = 3x - 2$
10) $y = \frac{5}{x}x - \frac{5}{4}$
11) $y = -x + 2$
12) $y = \frac{1}{3}x + 2$
13) $l = 34 - \frac{1}{2}d, 16\ days$
14) $28
15) 90

Point–slope form

1) $m = 3, (-3, 0)$
2) $m = 3, \left(\frac{2}{3}, 0\right)$
3) $m = 4, (-3, -9)$
4) $m = \frac{2}{3}, (0, -1)$
5) $m = 4, (-2, -3)$
6) $m = \frac{3}{2}, (0, -2)$
7) $m = 4, \left(\frac{1}{2}, 1\right)$
8) $m = \frac{2}{3}, \left(0, \frac{4}{3}\right)$
9) $m = 3, (1, -1)$
10) $m = 1, \left(-\frac{10}{4}, 2\right)$
11) $m = 32, (-1, -1)$
12) $m = 5, (-1, -2)$

13) $y = \frac{1}{2}x + 1$
14) $y = 3x - 5$
15) $y = -2x + 4$
16) $y = 3x - 11$
17) $y = -\frac{1}{2}x - 4$
18) $y = 3x - \frac{11}{4}$
19) $y = -\frac{3}{4}x - \frac{3}{2}$
20) $y = 2x + \frac{19}{14}$
21) $y = 5x + 2$
22) $y = \frac{4}{5}x + \frac{17}{5}$
23) $y = 4x + \frac{25}{6}$
24) $y = \frac{1}{5}x - 1$

25) $y = 10x - 37$
26) $y = -x - 11$
27) $y = -\frac{2}{3}x - 4$
28) $y = -4x + 5$
29) $y = -4x - 10$
30) $y = \frac{1}{2}x + \frac{1}{2}$
31) $y = \frac{7}{3}x + \frac{34}{3}$
32) $y = \frac{1}{2}x + 1$
33) $y = 2x - 7$
34) $y = -x$
35) $y = -2x - 2$
36) $y = 4x - 19$
37) $y = 3x - 5$

Equations of horizontal and vertical lines

1) y = 0 (it is on *x* axes)

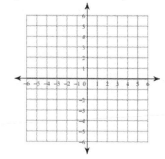

2) y = 2

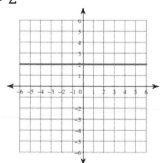

3) x = −4

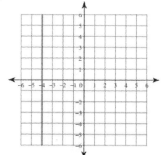

4) x = 3

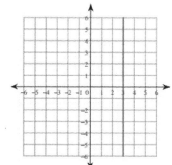

Equation of parallel or perpendicular lines

1) $y = \frac{3}{2}x + 6$

2) $y = -\frac{1}{3}x + 7$

3) $y = \frac{1}{3}x - 3$

4) $y = -2x$

5) $y = -\frac{1}{2}x + \frac{7}{2}$

6) $y = -\frac{1}{2}x + \frac{13}{2}$

7) $y = -x + 5$

8) $y = 3x - 4$

9) $y = x + 6$

10) $y = x + 5$

11) $y = -\frac{1}{3}x + \frac{1}{3}$

12) $y = \frac{1}{2}x - 7$

13) $y = -x$

14) $y = x + 6$

15) $y = \frac{1}{3}x - \frac{5}{3}$

16) $y = -\frac{1}{3}x + \frac{7}{3}$

17) $y = -2x - 6$

18) $y = \frac{1}{5}x + \frac{11}{5}$

19) $Y = 8x, \$32$

Chapter 9:

Monomials and Polynomials

Topics that you'll learn in this part:

- ✓ Classifying Polynomials
- ✓ Writing Polynomials in Standard Form
- ✓ Simplifying Polynomials
- ✓ Add and Subtract Monomials
- ✓ Multiplying Monomials
- ✓ Multiplying and Dividing Monomials
- ✓ GCF of Monomials
- ✓ Powers of Monomials
- ✓ Multiplying a Polynomial and a Monomial
- ✓ Multiplying Binomials
- ✓ Factoring Trinomials

Writing Polynomials in Standard Form

✍ Write each polynomial in standard form.

1) $3x^2 - 4x^3 + 1 =$
2) $7x - x^3 + 5x^2 =$
3) $15x + x^3 =$
4) $2(x^2 + 1) + x^4 =$
5) $(x - 1)x + 1) =$
6) $x(3x + 7) - x^4 =$
7) $15x - 4x^2 - 7x + 3 =$
8) $3(x - 2) - x^3 =$
9) $6x(x - 4) + 4x^2 + 1 =$
10) $3x^2 - 4x - 2(x^2 + 1) =$
11) $4x^3 + 2x - 5x^5 =$
12) $2 - 4x - 2(x^2 + 1) =$
13) $3x + (x + 1)(x + 1) =$
14) $4x - 5x^3 + 4 =$
15) $14y - 3y^2 + 2y =$
16) $4m(m^2 - 1) + 3m^4 =$
17) $2x - 4x^2 - 7x - 3x =$
18) $7x^3 + \frac{x^3 + x^2}{x^2} =$
19) $\frac{4N^4 + 4N}{2N} + N^2 =$
20) $17x^2 - 4x^5 + x^{10} =$
21) $-3x^5 + 4x - 9x^2 =$
22) $2x(x + 1) + \frac{3x + x^2}{x} =$
23) $17x - 4x^3 + 5x^2 + 1 =$

24) $(-3)\frac{x^2 - 9x + 15x^3}{3x} + x^3 =$
25) $14x^5 + \frac{x^5 + x^4}{2x^2} =$
26) $12x - 4 + 4x^3 + x^2 =$
27) $4x^3 + 7x^2 + 28x^5 =$
28) $34x^4 - 2x + 7x^2 =$
29) $2x^2(x^2 + 4x) + x =$
30) $-3x^2 + 4x - 7x^3 =$
31) $25x\left(\frac{1}{5}x^2 - \frac{x}{2}\right) =$
32) $x^2 + 4x^5 + x^3 =$
33) $2x^2 - 3x - 7x^2 - x =$
34) $2x^2 - 4x^4 + x^2 =$
35) $-3x + 7x^2 - x(x + 1) =$
36) $5x - x^2 + x^5 + x^3 =$
37) $4 + 2x - x^2 =$
38) $2z - 3 + z^3 =$
39) $p^6 - 2 + 3p^3 =$
40) $2y - 2y^6 + y^4 =$
41) $9x + x^2 - 7 =$
42) $2x + 1 =$
43) $3z^3 + 5 =$
44) $2 - 3q + q^2$

Simplifying Polynomials

✎ Simplify each expression.

1) $(-5)(4m^2 - 5m - 8)$
2) $2d(d^5 - 7d^3 + 4)$
3) $7rs(4r^2 + 9s^3 - 7rs)$
4) $8x^4(6x - 8x^3 + 7)$
5) $(y + 3)(y + 5)$
6) $(2x + 4)(x + 9)$
7) $(4b - 3)(4b + 3)$
8) $(n + 4m)(2n - 3m)$
9) $(x + 4)(6x^2 + 2x - 8)$
10) $(4z + 6)(4z + 6)$
11) $(x - 2)(9 - 5x)$
12) $(4b - 3)(4b - 3)$
13) $(3x^2 - 2x)(7x - 8x^2 - 9)$

✎ Solve.

14) Find the perimeter of the triangle pictured.

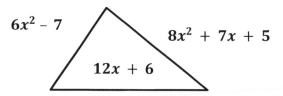

15) Find the area of the rectangles pictured.

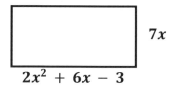

16) A triangle has sides of length $3x + 4y$, $5y + 6 + 2x$, and $7 + 8x$. What is the perimeter of the triangle?

17) Triangle has a perimeter $10x^2 - 3xy - 6y^2$. If two sides are known to be $2x^2 + 2xy$ and $7x^2 + 3y^2$, then what is the length of the third unknown side?

Add and Subtract monomials

✍ Find each sum and difference.

1) $3x^3y - 12x^3y =$

2) $10(3uv - u) + 5u =$

3) $12x^2z - z(5x^2 - z) =$

4) $3x^2(z + y) - 2x^2z =$

5) $14(x^2yz^2 - x) + x^2yz^2 + 12x =$

6) $6\frac{x^2}{y} - 4\frac{x^2}{y} =$

7) $3\left(x^2 + \frac{1}{y}\right) - x^2 + \frac{2}{y} =$

8) $3x^2z + (x^2)(2z) =$

9) $3\frac{z}{y}x^2 + 2x^2\left(1 + \frac{z}{y}\right) - x^2 =$

10) $3uvw - 15uvw =$

11) $12ux^2 - 14ux^2 =$

12) $3x(x^2 + y) + x^3 - 4xy =$

13) $3y^2zx^2 + 17x^2(zy^2 + 1) =$

14) $10m^6 + 12m^6 - 14m^6 =$

15) $32e^{-i\omega t} - 18e^{-i\omega t} =$

16) $15lq - 5lq + lq =$

17) $\frac{xyz}{2} + \frac{3}{2}xyz =$

18) $95\,lqr - 39lqr =$

19) $33\frac{z^3x}{y} - 19\frac{z^3x}{y} =$

20) $7(x^2z - y) + x^2z =$

21) $x^2zy - 2x^2(1 - zy) + 2x^2 =$

22) $33(x^2 + rs) - 30x^2 - 45rs =$

23) $12z^3x - 18z^3x =$

24) $x^2(yz) + 12x^2(yz) =$

25) $4xy - 15xy =$

26) $-2x^2y + x^2y - 3x^2y =$

27) $3\frac{x^2}{z} + 10\frac{x^2}{z} - 2\frac{x^2}{z} =$

28) $4yzu - 12yzu =$

29) $12ut - 14ut - 33ut =$

30) $15frq + 3frq =$

31) $12(x - yz) + 12x - 9yz =$

32) $10(x^2 - 2y) + 18y =$

33) $3\frac{xyz}{t} + 4\frac{xyz}{t} =$

34) $33(t^2e^{-i\omega t}) - (40t^2e^{-i\omega t}) =$

35) $\frac{t^3x}{z} + 4\left(\frac{t^3x}{z} - x\right) + 3x =$

36) $5wt - 12wt - 33wt =$

37) $2x + \frac{(x+1)^2}{x+1} =$

38) $3x^3 - 2x - 4x^3 + x =$

39) $2z^2 - 4z - 3z^2 =$

40) $\frac{(3x+1)(3x-1)}{3x-1} + x =$

41) $9r - 3r^2 - 8r =$

42) $3t(t + 1) - 3t =$

Multiplying Monomials

✍ Simplify each expression.

1) $(2x)(2x)^3 =$

2) $(2n^5)(n^2)^2 =$

3) $\left(\frac{x^2}{2}\right)^5 (3x) =$

4) $\left(\frac{2b}{3}\right)^4 (3b) =$

5) $-2x \left(\frac{1}{2x^2}\right)^4 =$

6) $(x\left(\frac{1}{2}x\right)^2)^3 =$

7) $5x^2z(2z) =$

8) $12xyz(x^2z) =$

9) $x^2y(12xz) =$

10) $3x^2z^3(2yzx) =$

11) $7(x^2zt)(t^2zx) =$

12) $(-2u^3vw)(4vw) =$

13) $(-5)(mn^3)(3nm^2)(mn) =$

14) $2(x^2y)(yz)(zxy) =$

15) $3v^2(9uw)(vw)^5 =$

16) $6(x^5)(xz^4)(zxy) =$

17) $(3c^3b)(b^2a)(cb) =$

18) $14x(3xz)(2yz) =$

19) $x^2(3yzx)(z^3xy) =$

20) $-5z^4(y^5z)(3xz) =$

21) $3(u^3v)(uv^3)(uvw)(wv) =$

22) $2(e^{-i\omega t}xt)(te^{-i\omega t}x) =$

23) $3(m^3n^2)(mn)(2n^3) =$

24) $(x^2z)(z^3xy)(yx) =$

25) $(g^2l)(l^2g)(gl) =$

26) $-2(x^2z)(3yx)(4zxy) =$

27) $(2x^5)(3xy^2)(zx) =$

28) $(p^3qt)(2qp)(t^2q) =$

29) $7(m^4nq)(2mn)(qmn^2) =$

30) $(12pt)(3p)(t^2p) =$

31) $(pls)(2p^2s)(s^3pl) =$

32) $\left(\frac{z^5}{t}\right)\left(3\frac{z^5}{t}\right)(-2)\left(\frac{z^5}{t}\right) =$

33) $(33x^2)(xy)(zxy) =$

34) $\left(2\left(\frac{z}{t}\right)^3\right)\left(4\frac{z^5}{t}\right)\left(\frac{z^5}{t}\right)^6 =$

35) $\left(\frac{2x}{z}\right)^5 \left(\frac{2x}{z}\right)\left(\frac{2x}{z}\right)^3 =$

36) $(\frac{m}{n^2})(2\left(\frac{m}{n^2}\right)^3) =$

37) $x^2(zy)(3y) =$

38) $4\left(\frac{x^2}{zy}\right)^2 \left(4\left(\frac{x^2}{zy}\right)\right) =$

39) $(\frac{\sqrt{a^2+b^2}}{b})(3\left(\frac{\sqrt{a^2+b^2}}{b}\right)^3) =$

40) $\left(3\left(\frac{t^4}{g}\right)\right)\left(2\left(\frac{t^4}{g}\right)^{10}\right) =$

Multiplying and Dividing Monomials

Simplify.

1) $\frac{3x^2yz^5}{xyz} =$

2) $\frac{25x(2xz^5)}{25x^2z^3} =$

3) $2\left(\frac{x^2m^5n^4}{x^3mn^4}\right)(2xn) =$

4) $3\frac{q}{r}\left(\frac{q^2r^5}{qr}\right) =$

5) $\frac{(3xy^3)(4xyz^3)}{12xyz} =$

6) $m^4(n^2)\left(\frac{5n^2}{m^3n^4}\right) =$

7) $\frac{3x(2xy)(3xz)}{9x^2} =$

8) $\left(\frac{3z^5x^2}{3z^3y}\right)\left(\frac{y}{z^3}\right) =$

9) $(3x^2y^2z^2)\left(\frac{2x}{x^2yz^2}\right) =$

10) $\left(\frac{e^{-i\omega t}}{t}\right)\left(\frac{t^3}{e^{-i\omega t}}\right) =$

11) $(3u^3v^2w)\left(\frac{1}{uvw^2}\right) =$

12) $2(mnl)(m^2nl^3)\left(\frac{1}{m^2l}\right) =$

13) $23(a^2b)\left(\frac{2ab}{b^2}\right) =$

14) $6(x^2zt)(xz^2t) =$

15) $\frac{3(ab)(c^2b)(abc)}{c^4a^3b} =$

16) $3x^2\left(\frac{(xz^3y)(2yz)}{x^2}\right)z^3 =$

17) $4x(12x^2)(z^4)\left(\frac{1}{z^3x^2}\right) =$

18) $\frac{(ab)(ab)^2(-2(ab)^5)}{(ab)^6} =$

19) $\left(\frac{x(w)^{2t}}{x(w)^t}\right)2xt =$

20) $2xt(t^5x)(2t^2) =$

21) $(\frac{(3a)(3b)(c^2ab)}{3(a^2bc^2)}) =$

22) $(-2x^2y)\left(\frac{1}{2zyx}\right) =$

23) $\frac{(x\sqrt{z})^5(2xz)(x\sqrt{z})}{2(x\sqrt{z})^3} =$

24) $3(x^2y^3)\left(\frac{2xy}{y^5zx^2}\right) =$

25) $(3uv^3)\left(\frac{uvw}{u^2vw}\right) =$

26) $(m^4n^6)(\frac{3nm(2n^2)}{3m^2n^3} =$

27) $x^2e^{3(i\omega t)}(\frac{15t}{e^{2(i\omega t)}x}) =$

28) $33xyz(2x^2zy) =$

29) $e^3hq\left(\frac{3hq}{6eq^2}\right) =$

30) $2(x^2z)(3zy)(zy) =$

31) $8z\left(\frac{2z^2-z^5}{8z^3}\right) =$

32) $\left(\frac{6q^4+3q^3}{3q}\right)\left(\frac{1}{q^2}\right) =$

33) $\left(\frac{1}{x^2-2x}\right)\left(\frac{10x^2-20x}{5}\right) =$

34) $(2x^2)\left(\frac{5}{x}\right) =$

GCF of Monomials

📝 *Find the GCF of each set of monomials.*

1) $6x^2z, 2xy$
2) $2x^2y^5z, 14yz$
3) $3u^3vw^4, 6vw$
4) $10x^2yz, 5y^2x$
5) $3x^2z^4, \left(\frac{x^2z^2}{x^2z}\right)$
6) $9uv^5w, 3uw$
7) $4trq^5, \left(\frac{t^3rq}{trq}\right)$
8) $\left(\frac{12x^2z^4y}{x^2zy}\right), -2xzy$
9) $11n^4pm^5, 11pmn^2$
10) $14xz^4ty^5, \left(\frac{7x^2yt^3}{xt}\right)$
11) $6x^2y^3, xyz, 3x^2$
12) $15d^3eh^3, 5edh, 7deh^2$
13) $\left(\frac{2x^2y^5z}{yz}\right), 3xz, 4x^2y$
14) $33x^5zy^5, 22x^2y$
15) $\left(\frac{(-2xyz)(x^2)}{3x^3z}\right), 2xzy$
16) $21x^2mn, 35nx$
17) $3a(b^2ac), a^2c, 3ab$
18) $\left(\frac{m^4n^2p}{np}\right), 2m(nm)$
19) $15x^2\left(\frac{3x^2yz}{xz}\right), 5y^5zx^2$

20) $e^5hq, \left(\frac{3e^6h^4q}{e^4hq}\right)$
21) $\left(\frac{2x^2(4z^4x^2y)}{4xz^2y}\right), 32x^2yz$
22) $14mp^5t^4, 7m^6t^3p^2$
23) $\left(\frac{(-2mn)(n^4m^3)}{n^5m}\right), -n^5m^7$
24) $(x^2)(-7yz^5x^2), 2x^2y$
25) $25\left(\frac{x^2yz^4}{15yz}\right), 3x^2yz^8$
26) $(-3ab)(3bc), 3acb^2$
27) $\left(\frac{(ab)(bc)(ca)}{2ba^2}\right), 8a^3b^6c$
28) $9r^3q^5s^2, \left(\frac{(rsq^2)(3sq)}{3srq}\right)$
29) $2mny^3t, 4mty$
30) $3(abc)(2ca), 20a^2cb$
31) $2\frac{(x^2y^5z^4)}{2xz}, 2xy$
32) $3xy(2xyz), x^2z(yx)$
33) $25azx(3xz), 25a^3zx^2$
34) $33x^2 + 11x, 11x^2 + 22x$
35) $\frac{5}{(x-1)(x+3)}, \frac{-5}{x-1}$
36) $\frac{2x^5+x^2}{x}, x^2$
37) $2z^5 + z, z^3$
38) $2x(1-y), (x - xy)$

Powers of monomials

Simplify.

1) $\left(\frac{1}{2x}\right)\left(\frac{2x}{3}\right)^3 =$

2) $\left(\frac{3xy}{3z}\right)^2 (2z) =$

3) $\left(\frac{x^2zt}{2z^3}\right)^3 (8z^9) =$

4) $\left(\frac{(2x)^2}{yz}\right)^3 =$

5) $\left(\frac{(2m+1)}{3m^2}\right)^2 =$

6) $(zt^{10}b)^2(3z) =$

7) $(2x^2yz)^3 =$

8) $3(2mn^2)^4 =$

9) $(2x^2z^4y)^2 =$

10) $4(y^3xz)^5 =$

11) $(x^2y^2z^4)^3 =$

12) $(-2x^{ab}y^3)^3 =$

13) $\left(\frac{x^2y}{z}\right)^{ab} =$

14) $(-2a^6b^2c^4)^{5x} =$

15) $(x^2y^{2ab}z^{2b})^3 =$

16) $\left(\frac{x^2yz^5}{t}\right)^6 =$

17) $\left(\frac{(3x^2y^3z)}{2zy}\right)^2 =$

18) $(4x^2z^4)^{ab} =$

19) $(m^{10}n^6r)^3 =$

20) $(2yx^2z)^2 =$

21) $(x^2y^2z^3)^{2t} =$

22) $(3x^2yz^3)^5 =$

23) $\left(\frac{(2xyz^2)^4}{2xy}\right)^2 =$

24) $\left(\frac{(x^2yz^3)^3}{(xz)^2}\right)^2 =$

25) $\frac{(2xy)^3}{(x^2yz^4)^b} =$

26) $(x^2z)^2(2xy)^4 =$

27) $3(2x^2z)^5 =$

28) $(a^4bc^5)^{12} =$

29) $(10^x 12^y 5^z)^m =$

30) $(\propto^5 \beta^6 \gamma^8)^2 =$

31) $\left(\frac{(x^2y)^2(2zy)^2}{(xyz)^a}\right)^2 =$

32) $(2xy^5z^4x^2)^3 =$

33) $\left(\frac{x^2yz^2}{(xy)^a}\right)^b =$

34) $(t^6r^5s^6)^{(a+b)} =$

35) $2(x^2yx^4)^3 =$

36) $(abc^2)^2(a^2)^4 =$

37) $2(2x)^2(x^2yz)^2 =$

38) $\left(\frac{(xy^3z^2)^2}{x^2(2yz)^3}\right) =$

39) $(2x)^2(x^6z^3)^{\frac{1}{3}} =$

40) $(t^2x^2y^5)^{3ab} =$

Multiplying a Polynomial and a Monomial

✏️ *Find each product.*

1) $(2xy)\left(\frac{1}{2x+1}\right) =$

2) $\left(\frac{1}{3z+4}\right)\left(\frac{2xy+1}{3}\right) =$

3) $(-2rq)(3q - 3r + 1) =$

4) $\frac{3ca^3}{2}\left(1 + \frac{2}{ca}\right) =$

5) $3x(2x + 1) =$

6) $3b(b^2 - 1) =$

7) $2x(1 - 2x) =$

8) $\frac{x}{3}(3x^2 - 6y^3) =$

9) $yx^2(z^2 + xy) =$

10) $2x^4(z^2 - y) =$

11) $3ab(a^2 - b^2) =$

12) $z^2 t\left(3ztx - \frac{2}{zt}\right) =$

13) $\frac{x}{y}\left(2x^2 + 3y - \frac{y}{x}\right) =$

14) $3ac(a^2 - 2ab + b^2) =$

15) $(-b)\left(\frac{3x}{2} + \frac{2x}{b}\right) =$

16) $a^b(x^2 - 2x + 4) =$

17) $(-2xy)(c - b + a) =$

18) $p\left(\frac{l}{s} - 2ls - s^2\right) =$

19) $2x^2 yz(1 - 4x - 2y) =$

20) $y^3 xz\left(25x - \frac{1}{y^3}\right) =$

21) $3uv(u^2 + 2uv + v^2) =$

22) $\left(-2\frac{x}{yz}\right)\left(\frac{2yz^2}{3x} - 2yz - xyz\right) =$

23) $3b^2(b^2 - 4ac) =$

24) $\frac{-b}{2a}(1 - b^2 - ac) =$

25) $5xy(3zx - xy + 2x^2) =$

26) $x^2 y(3 - z) =$

27) $(2x + 2y - z)(-2y) =$

28) $2a(-x^2 - 2y^3 + xy) =$

29) $\frac{3xz}{y^3}\left(1 - \frac{2y^3}{xz}\right) =$

30) $7v(2ut - 2vt) =$

31) $4u^4(1 - 2v + 5w) =$

32) $\frac{x^2}{3z}\left(z^2 + 2z - \frac{2}{x^2}\right) =$

33) $6mz\left(\frac{12+m}{3z}\right) =$

34) $e^{2x}(1 - 3xz + 4y) =$

35) $\frac{2x^2}{z}(3 + 2z - y) =$

36) $(-x)^2(x^2 + 2x - 1) =$

37) $2v^2 w\left(\frac{1-3u+5vw}{vw} - 1\right) =$

38) $2x^0(3x + 4z - 2zxy) =$

39) $4x(2x - 1) =$

40) $18x^2\left(\frac{1}{3x^2}\right) =$

Multiplying Binomials

📝 *Simplify each expression.*

1) $(x-5)(x+4) =$

2) $(x-6)(x-3) =$

3) $(x+4)(x+7) =$

4) $(x+3)(x-7) =$

5) $(3x-5)(2x+8) =$

6) $(11x-7)(5x+3) =$

7) $(4x-9)(9x+4) =$

8) $(x-2)(x+2) =$

9) $(x-2)(x-2) =$

10) $(x+4)(x-5) =$

11) $(3x-7)(x+3) =$

12) $(x-5)(4x+9) =$

13) $(3x+7)(8x-1) =$

14) $(x-1)(x+4) =$

15) $(10x+7)(x-11) =$

16) $(4x^2+1)(3-x) =$

📝 *Find the area of each shape.*

17)

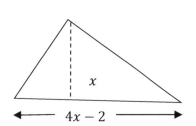

19)

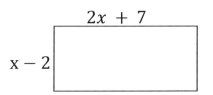

18)

20)

Factoring Trinomials

✍ Factor each trinomial.

1) $x^2 + 8x + 15 =$

2) $x^2 - 5x + 6 =$

3) $x^2 + 6x + 8 =$

4) $x^2 - 6x + 8 =$

5) $x^2 - 8x + 16 =$

6) $x^2 - 7x + 12 =$

7) $x^2 + 11x + 18 =$

8) $x^2 + 2x - 24 =$

9) $x^2 + 4x - 12 =$

10) $x^2 - 10x + 9 =$

11) $x^2 + 5x - 14 =$

12) $x^2 - 6x - 27 =$

13) $x^2 - 11x - 42 =$

14) $x^2 + 22x + 121 =$

15) $6x^2 + x - 12 =$

16) $x^2 - 17x + 30 =$

17) $3x^2 + 11x - 4 =$

18) $10x^2 + 33x - 7 =$

19) $x^2 + 24x + 144 =$

20) $8x^2 + 10x - 3 =$

✍ Solve.

21) A certain company's main source of income is a mobile app. The company's annual profit (in millions of dollars) as a function of the app's price (in dollars) is modeled by $P(x) = -2(x - 3)(x - 11)$. Which app prices will result in $0 annual profit?

22) A rectangular plot is 6 meters longer than it is wide. The area of the plot is 16 square meters. Find the length and width of the plot.

23) The combined area of two squares is 20 square centimeters. Each side of one square is twice as long as a side of the other square. Find the lengths of the sides of each square.

Answers of Worksheets

Writing Polynomials in Standard Form

1) $-4x^3 + 3x^2 + 1$
2) $-x^3 + 5x^2 + 7x$
3) $x^3 + 15$
4) $x^4 + 2x^2 + 2$
5) $x^2 - 1$
6) $-x^4 + 3x^2 + 7x$
7) $-4x^2 + 8x + 3$
8) $-x^3 + 3x - 6$
9) $10x^2 - 24x + 1$
10) $x^2 - 4x - 2$
11) $-5x^5 + 4x^3 + 2x$
12) $-2x^2 - 4x$
13) $x^2 + 5x + 1$
14) $-5x^3 + 4x + 4$
15) $-3y^2 + 16y$
16) $3m^4 + 4m^3 - 4m$
17) $-4x^2 - 8x$
18) $7x^3 + x + 1$
19) $2N^3 + N^2 + 2$
20) $x^{10} - 4x^5 + 17x^2$
21) $-3x^5 - 9x^2 + 4x$
22) $2x^2 + 3x + 3$
23) $-4x^3 + 5x^2 + 17x + 1$
24) $x^3 - 15x^2 - x + 9$
25) $14x^5 + \frac{1}{2}x^3 + \frac{1}{2}x^2$
26) $4x^3 + x^2 + 12x - 4$
27) $28x^5 + 4x^3 + 7x^2$
28) $34x^4 + 7x^2 - 2x$
29) $2x^4 + 8x^3 + x$
30) $-7x^3 - 3x^2 + 4x$
31) $5x^3 - \frac{25}{2}x^2$
32) $4x^5 + x^3 + x^2$
33) $-5x^2 - 4x$
34) $-4x^4 + 3x^2$
35) $6x^2 - 4x$
36) $x^5 + x^3 - x^2 + 5x$
37) $-x^2 + 2x + 4$
38) $z^3 + 2z - 3$
39) $p^6 + 3p^3 - 2$
40) $-2y^6 + y^4 + 2y$
41) $x^2 + 9x - 7$
42) $2x + 1$
43) $3z^3 + 5$
44) $q^2 - 3q + 2$

Simplifying Polynomials

1) $-20m^2 + 25m + 40$
2) $2d^6 - 14d^4 + 8d$
3) $28r^3s + 63rs^4 - 49r^2s^2$
4) $-64x^7 + 48x^5 + 56x^4$

5) $y^2 + 8y + 15$
6) $2x^2 + 22x + 36$
7) $16b^2 - 9$
8) $2n^2 - 12m^2 + 5mn$
9) $6x^3 + 26x^2 - 32$
10) $16z^2 + 48z + 36$
11) $-5x^2 + 19x - 18$
12) $16b^2 - 24b + 9$
13) $-24x^4 + 37x^3 - 41x^2 - 18x$
14) $14x^2 + 19x - 1$
15) $14x^3 + 42x^2$
16) $13x + 9y + 13$
17) $x^2 - 5xy - 9y^2$

Add and Subtract Monomials

1) $-9x^3y$
2) $(30uv - 5u)$
3) $7x^2z + z^2$
4) $3x^2y + x^2z$
5) $15x^2yz^2 - 2x$
6) $2\frac{x^2}{y}$
7) $2x^2 + \frac{1}{y}$
8) $5x^2z$
9) $5\frac{z}{y}x^2 + x^2$
10) $-12uvw$
11) $-2ux^2$
12) $4x^3 - xy$
13) $20y^2zx^2 + 17x^2$
14) $8m^6$
15) $14e^{-i\omega t}$
16) $11lq$
17) $2xyz$
18) $56\ lqr$
19) $14\frac{z^3x}{y}$
20) $8x^2z - 7y$
21) $3x^2zy$
22) $3x^2 - 12rs$
23) $-6z^3x$
24) $13x^2yx$
25) $-11xy$
26) $-4x^2y$
27) $11\frac{x^2}{z}$
28) $-8yzu$
29) $-35ut$
30) $18frq$
31) $24x - 21yz$
32) $10x^2 - 2y$
33) $7\frac{xyz}{t}$
34) $-7(t^2e^{-i\omega t})$
35) $5\frac{t^3x}{z} - x$
36) $-40wt$
37) $3x + 1$
38) $-x^3 - x$
39) $-z^2 - 4z$
40) $4x + 1$
41) $-3r^2 + r$
42) $3t^2$

Multiplying Monomials

1) $(2x)^4$
2) $2n^9$
3) $\frac{3x^{11}}{32}$
4) $\frac{16b^5}{27}$
5) $\frac{-1}{8x^7}$
6) $\frac{x^9}{64}$
7) $10x^2z^2$
8) $12x^3yz^2$
9) $12x^3yz$
10) $6x^3z^4y$
11) $7(x^3z^2t^3)$
12) $(-8u^3v^2w^2)$
13) $(-15)(m^4n^5)$
14) $2(x^3y^3z^2)$
15) $27v^7uw^6$

16) $6(x^7z^5y)$
17) $(3ac^4b^4)$
18) $84x^2yz^2$
19) $3x^4y^2z^4$
20) $-15xy^5z^6$
21) $3(u^5v^6w^2)$
22) $2(e^{-2i\omega t}x^2t)$
23) $6(m^4n^6)$
24) $(x^4z^4y^2)$
25) (g^4l^4)
26) $-24(x^4y^2z^2)$
27) $(6x^7y^2z)$
28) $(2p^4q^3t^3)$
29) $14(m^6n^4q^2)$
30) $(36p^3t^3)$
31) $2(p^4l^2s^5)$
32) $(-6)\frac{z^{15}}{t^3}$
33) $(33x^4y^2z)$
34) $8\frac{z^{38}}{t^{10}}$
35) $\frac{512x^9}{z^9}$
36) $2\left(\frac{m}{n^2}\right)^4$
37) $3x^2y^2z$
38) $16\left(\frac{x^2}{zy}\right)^3$
39) $3\left(\frac{\sqrt{a^2+b^2}}{b}\right)^4$
40) $6\left(\frac{t^4}{g}\right)^{11}$

Multiplying and Dividing Monomials

1) $3xz^4$
2) $2z^2$
3) $4m^4n$
4) $3q^2r^3$
5) xy^3z^2
6) $5m$
7) $2xyz$
8) $\frac{x^2}{z}$
9) $6xy$
10) t^2
11) $\left(\frac{3u^2v}{w}\right)$
12) $2(mn^2l^3)$
13) $46a^3$
14) $6(x^2z^3t^2)$
15) $\frac{3b^2}{ca}$
16) $6z^7xy^2$
17) $48xz$
18) $-2(ab)^2$
19) $2xwt$
20) $4x^2t^8$
21) $3b$
22) $\left(-\frac{x}{z}\right)$
23) $xz(x\sqrt{z})^3$
24) $\frac{6x}{yz}$
25) $3v^3$
26) $2m^3n^6$
27) $15txe^{(i\omega t)}$
28) $66x^3y^2z^2$
29) $\frac{1}{2}e^2h^2$
30) $6(x^2z^3y^2)$
31) $2-z^3$
32) $2q+1$
33) 2
34) $10x$

GCF of monomials

1) $2x$
2) $2yz$
3) $3vw$
4) $5xy$
5) z
6) $3uw$
7) t
8) $2z$
9) $11pmn^2$
10) $7xt$
11) x
12) deh
13) x
14) $11x^2y$
15) $2y$
16) xn
17) a
18) m^2n
19) $5x^2y$
20) e^2h

21) $2x^2z$ 26) $3acb^2$ 31) $2xy$ 36) x

22) $7mp^2t^3$ 27) bc 32) x^2yz 37) z

23) $-m^3$ 28) q^2s 33) $25azx^2$ 38) $x(1-y)$

24) x^2y 29) $2mty$ 34) $11x$

25) x^2z^3 30) $2a^2cb$ 35) $\frac{5}{x-1}$

Powers of monomials

1) $\frac{4x^2}{27}$

2) $\frac{2x^2y^2}{z}$

3) $x^6z^3t^3$

4) $\frac{64x^6}{y^3z^3}$

5) $\frac{(2m+1)^2}{9m^4}$

6) $3z^3t^{20}b^2$

7) $8x^6y^3z^3$

8) $48m^4n^8$

9) $4x^4z^8y^2$

10) $4y^{15}x^5z^5$

11) $x^6y^6z^{12}$

12) $-8x^{3ab}y^9$

13) $\frac{x^{2ab}y^{ab}}{z^{ab}}$

14) $(-2)^{5x}a^{30x}b^{10x}c^{20x}$

15) $x^6y^{6ab}z^{6b}$

16) $\frac{x^{12}y^6z^{30}}{t^6}$

17) $\frac{9x^4y^6z^2}{4z^2y^2}$

18) $4^{ab}x^{2ab}z^{4ab}$

19) $m^{30}n^{18}r^3$

20) $4y^2x^4z^2$

21) $x^{4t}y^{4t}z^{6t}$

22) $243x^{10}y^5z^{15}$

23) $\frac{256x^8y^8z^{16}}{4x^2y^2}$

24) $\frac{x^{12}y^6z^{18}}{x^4z^4}$

25) $\frac{8x^3y^3}{x^{2b}y^bz^{4b}}$

26) $16x^8z^2y^4$

27) $96x^{10}z^5$

28) $a^{48}b^{12}c^{60}$

29) $10^{mx}12^{my}5^{mz}$

30) $\alpha^{10}\beta^{12}\gamma^{16}$

31) $\frac{16x^8y^8z^4}{x^{2a}y^{2a}z^{2a}}$

32) $8x^9y^{15}z^{12}$

33) $\frac{x^{2b}y^bz^{2b}}{x^{ab}y^{ab}}$

34) $t^{6(a+b)}r^{5(a+b)}s^{6(a+b)}$

35) $2x^6y^3x^{12}$

36) $a^{10}b^2c^4$

37) $8x^6y^2z^2$

38) $\frac{x^2y^6z^4}{8x^2y^3z^3}$

39) $4x^4z$

40) $t^{6ab}x^{6ab}y^{15ab}$

Multiplying a Polynomial and a Monomial

1) $\frac{2xy}{2x+1}$

2) $\frac{2xy+1}{9z+12}$

3) $-6rq^2+6r^2q-2rq$

4) $\frac{3ca^3}{2}+3a^2$

5) $6x^2+3x$

6) $3b^3-3b$

7) $2x-4x^2$

8) x^3-2xy^3

9) $yx^2z^2+x^3y^2$

10) $2x^4z^2-2x^4y$

11) $3a^3b - 3ab^3$

12) $3z^3t^2x - 2z$

13) $\frac{2x^3}{y} + 3x - 1$

14) $3ca^3 - 6a^2bc + 3acb^2$

15) $-\frac{3bx}{2} - 2x$

16) $x^2a^b - 2xa^b + 4a^b$

17) $-2xyc + 2xyb - 2xya$

18) $\frac{\rho l}{s} - 2\rho ls - \rho s^2$

19) $(2x^2yz - 82x^3yz - 4x^2y^2z)$

20) $25y^3x^2z - xz$

21) $3u^3v + 6u^2v^2 + 3uv^3$

22) $\frac{-4z}{3} + 4x + 2x^2$

23) $3b^4 - 12ab^2c$

24) $\frac{-b}{2a} + \frac{b^3}{2a} + \frac{bc}{2}$

25) $15x^2yz - 5x^2y^2 + 10x^3y$

26) $3x^2y - x^2yz$

27) $-4xy - 4y^2 + 2yz$

28) $-2ax^2 - 4ay^3 + 2axy$

29) $\frac{3xz}{y^3} - 6$

30) $14uvt - 14v^2t$

31) $u^4 - 8u^4v + 20u^4w$

32) $\frac{x^2z}{3} + \frac{2x^2}{3} - \frac{2}{3z}$

33) $24m + 2m^2$

34) $e^{2x} - 3xze^{2x} - 4ye^{2x}$

35) $\frac{6x^2}{z} + 4x^2 - \frac{2x^2y}{z}$

36) $x^4 + 2x^3 - x^2$

37) $2v - 6uv + 10v^2w - 2v^2w$

38) $(6x + 8z - 4zxy)$

39) $8x^2 - 4x$

40) 6

Multiplying Binomials

1) $x^2 - x - 20$

2) $x^2 - 9x + 18$

3) $x^2 + 11x + 28$

4) $x^2 - 4x - 21$

5) $6x^2 + 14x - 40$

6) $55x^2 - 2x - 21$

7) $36x^2 - 65x - 36$

8) $x^2 - 4$

9) $x^2 - 4x + 4$

10) $x^2 - x - 20$

11) $3x^2 + 2x - 21$

12) $4x^2 - 11x - 45$

13) $24x^2 + 53x - 7$

14) $x^2 + 3x - 4$

15) $10x^2 - 103x - 77$

16) $-4x^3 + 12x^2 - x + 3$

17) $x^2 + 6x + 9$

18) $x^2 - x$

19) $x^2 + 9x + 18$

20) $2x^2 + 3x - 14$

Factoring Trinomials

1) $(x+3)(x+5)$
2) $(x-2)(x-3)$
3) $(x+4)(x+2)$
4) $(x-2)(x-4)$
5) $(x-4)(x-4)$
6) $(x-3)(x-4)$
7) $(x+2)(x+9)$
8) $(x+6)(x-4)$
9) $(x-2)(x+6)$
10) $(x-1)(x-9)$
11) $(x-2)(x+7)$
12) $(x-9)(x+3)$
13) $(x+3)(x-14)$
14) $(x+11)(x+11)$
15) $(2x+3)(3x-4)$
16) $(x-15)(x-2)$
17) $(3x-1)(x+4)$
18) $(5x-1)(2x+7)$
19) $(x+12)(x+12)$
20) $(4x-1)(2x+3)$
21) $3, 11$
22) $w=2, l=8$
23) $s=2$

Chapter 10:

Exponents and Radicals

Topics that you'll learn in this part:

✓ Multiplication Property of Exponents

✓ Division Property of Exponents

✓ Powers of Products and Quotients

✓ Zero and Negative Exponents

✓ Negative Exponents and Negative Bases

✓ Writing Scientific Notation

✓ Square Roots

Multiplication Property of Exponents

🖉 Simplify

1) $4x^4 \times 4x^4 \times 4x^4 =$

2) $2x^2 \times x^2 =$

3) $x^4 \times 3x =$

4) $x \times 2x^2 =$

5) $5x^4 \times 5x^4 =$

6) $2yx^2 \times 2x =$

7) $3x^4 \times y^2x^4 =$

8) $y^2x^3 \times y^5x^2 =$

9) $4yx^3 \times 2x^2y^3 =$

10) $6x^2 \times 6x^3y^4 =$

11) $3x^4y^5 \times 7x^2y^3 =$

12) $7x^2y^5 \times 9xy^3 =$

13) $7xy^4 \times 4x^3y^3 =$

14) $3x^5y^3 \times 8x^2y^3 =$

15) $3x \times y^5x^3 \times y^4 =$

16) $yx^2 \times 2y^2x^2 \times 2xy =$

🖉 Solve.

17) There are 7^6 pieces of leaves on a tree, and there are 7^4 trees in a forest. How many pieces of leaves are there in the forest?

18) In a storage warehouse, each container weights 6^3 pounds. If there are 6^5 containers, how much do the crates weigh in total?

19) You own a microscope with an objective lens and an eyepiece. The objective lens can magnify an object 10^3 times, and the eyepiece can further magnify an object 10^2 times. What is the maximum magnification on your microscope?

20) An asteroid travel at a speed of 8^8 miles per day, how many miles will it travel in 8^3 day?

Division Property of Exponents

Simplify

1) $\dfrac{3x^3}{2x^5}$

2) $\dfrac{12x^3}{14x^6}$

3) $\dfrac{12x^3}{9y^8}$

4) $\dfrac{25xy^4}{5x^6y^2}$

5) $\dfrac{2x^4}{7x}$

6) $\dfrac{16x^2y^8}{4x^3}$

7) $\dfrac{12x^4}{15x^7y^9}$

8) $\dfrac{12yx^4}{10yx^8}$

9) $\dfrac{16x^4y}{9x^8y^2}$

Solve.

10) Dalloway's room has the dimensions $3a^7$ by, $4b^3$ by, $5b^2$. What is the volume of Dalloway's room?

11) The fuel tank of Mr. Lee's car has the dimensions b^5 by $3c^2$ by $2c^3$. What is the volume of the fuel tank?

12) A factory produces wardrobes and likes to use exponents as dimensions. The wardrobes have the dimensions b^4 by b^4 by $5c^6$. What is the volume of the wardrobes?

13) The annual corn yield is $5a^2 kg$ per hectare. If there are $2b^8$ hectares of corn field in Nebraska and $7b^7$ hectares of corn field in Illinois, what is the total annual corn yield in these two states?

14) The dimensions of a water tank are a^2mm by a^5mm by $3b^2$mm. If 1ml water may contain $3c^{21}$ water molecules, how many water molecules are there in the water tank?

Powers of Products and Quotients

✍ Simplify.

1) $(2yx^2)^3 =$
2) $(2z^3y^3x)^4 =$
3) $\left(\frac{x^2yz}{4}\right)^3 =$
4) $(a^{10}b^3c)^2 =$
5) $(2x^2zy^5)^3 =$
6) $\left(\frac{2x}{z^2}y^4\right)^5 =$
7) $(2r^5q^2t^2)^3 =$
8) $\left(\frac{3x^2yz^4}{2m^4n}\right)^3 =$
9) $(10^2 \cdot 10^3)^8 =$
10) $(3^a \cdot 3^b)^4 =$
11) $\left(2x^3y^{10}\frac{z}{3}\right)^3 =$
12) $(xz^4y^{10})^5 =$
13) $\left(\frac{m^6n^4t^3}{2x^2}\right)^5 =$
14) $\left(\frac{i^3j^5k^3}{2ik}\right)^5 =$
15) $(6^5 \cdot 6^2 \cdot 6^0)^2 =$
16) $(e^2 \cdot e^4)^4 =$
17) $\left(t^2\frac{p^2}{q^4}\right)^6 =$
18) $\left(y^5x^4\frac{1}{z^6}\right)^3 =$
19) $\left(\frac{2x \cdot 3y}{3z \cdot x^2}\right)^3 =$
20) $(3xy \cdot 4xy)^3 =$

21) $(-j^3 \cdot 2j)^3 =$
22) $\left(\frac{-2x^2y^2}{z}\right)^3 =$
23) $(-3e^{2t}e^{2\omega t})^3 =$
24) $\left(\frac{25x^{10}y^8}{3z^9}\right)^0 =$
25) $\left(\frac{(-2x)^2}{(-2yz)^3}\right)^3 =$
26) $(4x^2y^4)^4 =$
27) $(2x^4y^4)^3 =$
28) $(3x^2y^2)^2 =$
29) $(3x^4y^3)^4 =$
30) $(2x^6y^8)^2 =$
31) $(12x^3x)^3 =$
32) $(2x^9x^6)^3 =$
33) $(5x^{10}y^3)^3 =$
34) $(4x^3x^3)^2 =$
35) $(3x^3 \cdot 5x)^2 =$
36) $(10x^{11}y^3)^2 =$
37) $(9x^7y^5)^2 =$
38) $(4x^4y^6)^5 =$
39) $(3x \cdot 4y^3)^2 =$
40) $\left(\frac{5x}{x^2}\right)^2 =$
41) $\left(\frac{x^4y^4}{x^2y^2}\right)^3 =$
42) $\left(\frac{25x}{5x^6}\right)^2 =$

43) $\left(\frac{x^8}{x^6y^2}\right)^2 =$
44) $\left(\frac{xy^2}{x^3y^3}\right)^{-2} =$
45) $\left(\frac{2xy^4}{x^3}\right)^2 =$
46) $\left(\frac{xy^4}{5xy^2}\right)^{-3} =$
47) $((3xyz)^2)^{\frac{1}{2}} =$
48) $(x^2y^2)^{\frac{1}{2}} =$
49) $3x(w^3)^3 =$
50) $\left(\frac{2x}{(2-x)}\right)^2 =$
51) $\left(\frac{2}{3x^2y}\right)^3 (y^2x) =$
52) $\frac{(2rq^2)^2}{(-q)^3} =$
53) $\left(\frac{-x^4}{3zy}\right)^2 =$
54) $(2x \cdot x \cdot x^2)^3 =$
55) $(-xy)^2 =$
56) $\frac{(2x^2y)^4}{4x^3} =$
57) $(3x^2)^{t+1} =$
58) $\left(\frac{2tx^2}{3xt^3}\right)^m =$
59) $(3xyz^5)^r =$

Zero and Negative Exponents

✎ Evaluate the following expressions.

1) $8^{-1} =$
2) $8^{-2} =$
3) $2^{-4} =$
4) $10^{-2} =$
5) $9^{-1} =$
6) $3^{-2} =$
7) $7^{-2} =$
8) $3^{-4} =$
9) $6^{-2} =$
10) $5^{-3} =$
11) $22^{-1} =$
12) $4^{-2} =$
13) $5^{-2} =$
14) $35^{-1} =$
15) $4^{-3} =$
16) $6^{-3} =$
17) $3^{-5} =$
18) $5^{-2} =$
19) $2^{-3} =$
20) $3^{-3} =$
21) $7^{-3} =$
22) $6^{-3} =$
23) $8^{-3} =$
24) $9^{-2} =$

25) $10^{-3} =$
26) $10^{-9} =$
27) $(\frac{1}{2})^{-1} =$
28) $(\frac{1}{2})^{-2} =$
29) $(\frac{1}{3})^{-2} =$
30) $(\frac{2}{3})^{-2} =$
31) $(\frac{1}{5})^{-3} =$
32) $(\frac{3}{4})^{-2} =$
33) $(\frac{2}{5})^{-2} =$
34) $(\frac{1}{2})^{-8} =$
35) $(\frac{2}{5})^{-3} =$
36) $(\frac{3}{7})^{-2} =$
37) $(\frac{5}{6})^{-3} =$
38) $(\frac{x^2}{e^{-2t}})^{-2} =$
39) $(3xz^{-3})^2 =$
40) $(\frac{(a^3)^{-2}}{b})^{-2} =$
41) $(\frac{x^2 y}{(-2z)^2})^{-3} =$
42) $(\frac{gh^3}{2k})^{-3} =$
43) $(\frac{2sqr^2}{2x})^{-1} =$

44) $(\frac{25xyz}{33mn^3})^0 =$
45) $(2xy(x^2)^{-2})^{-1} =$
46) $(x^2 z^3)^{-3} =$
47) $(\frac{1}{3}xy^2)^{-2} =$
48) $\frac{(y^{20}x^{15})^0}{(2x)^{-2}} =$
49) $(-3xy)^{-2} =$
50) $(e^{-\omega t})(e^{\omega t}) =$
51) $(3x + y)^{-2} =$
52) $\frac{(4x+1)^{-1}}{(9x)^{-3}} =$
53) $(\frac{2y}{zx})^{-3} =$
54) $(\frac{1}{2})^{-5} =$
55) $(\frac{3}{(2-2x^2)})^{-2} =$
56) $(\frac{1}{2x+1})^{-1}(2x+1)^{-1} =$
57) $(1-x)^{-2} =$
58) $(x^2)^{-2} =$
59) $(x^{-2}zy)^{-2} =$
60) $(x^{-1}z)^2 =$
61) $(yz^{-2})^{-1} =$
62) $(2x)^{-3}(x^2) =$

Writing Scientific Notation

✍ **Write each number in scientific notation.**

1) $0.113 = 1.13 \times 10^{-1}$
2) $0.02 = 2 \times 10^{-2}$
3) $2.5 = 2.5$
4) $20 = 2 \times 10^{1}$
5) $60 = 6 \times 10^{1}$
6) $0.004 = 4 \times 10^{-3}$
7) $78 = 7.8 \times 10^{1}$

8) $1{,}600 = 1.6 \times 10^{3}$
9) $1{,}450 = 1.45 \times 10^{3}$
10) $91{,}000 = 9.1 \times 10^{4}$
11) $2{,}000{,}000 = 2 \times 10^{6}$
12) $0.0000006 = 6 \times 10^{-7}$
13) $354{,}000 = 3.54 \times 10^{5}$
14) $0.000325 = 3.25 \times 10^{-4}$

15) $0.00023 = 2.3 \times 10^{-4}$
16) $56{,}000{,}000 = 5.6 \times 10^{7}$
17) $21{,}000 = 2.1 \times 10^{4}$
18) $78{,}000{,}000 = 7.8 \times 10^{7}$
19) $0.0000022 = 2.2 \times 10^{-6}$
20) $0.00012 = 1.2 \times 10^{-4}$
21) $0.02 = 2 \times 10^{-2}$

✍ **Solve.**

22) A color photograph taken with a digital camera is converted into digital format using 4×10^0 bytes per pixel. Photographs taken with the camera each have 2.2×10^6 pixels. How many bytes are there in one photo? Write your answer in scientific notation. 8.8×10^6

23) A certain animated movie earned 1.1×10^9 in revenues at the box office. The movie lasts $\$9.1 \times 10^1$ minute. How much revenue was earned per minute of the movie? Write your final answer in scientific notation 1.21×10^7

24) The weight of a honeybee is $1.2 \times 10^{-1} g$. The weight of the pollen collected by the bee on one trip is $6.2 \times 10^{-2} g$. What is the combined weight of the bee and the pollen? Express your answer in scientific notation. 1.8×10^{-1}

Square Roots

✎ Find the value each square root.

1) $\sqrt{1} =$
2) $\sqrt{4} =$
3) $\sqrt{9} =$
4) $\sqrt{25} =$
5) $\sqrt{16} =$
6) $\sqrt{49} =$
7) $\sqrt{36} =$

8) $\sqrt{0} =$
9) $\sqrt{64} =$
10) $\sqrt{81} =$
11) $\sqrt{121} =$
12) $\sqrt{225} =$
13) $\sqrt{144} =$
14) $\sqrt{100} =$

15) $\sqrt{256} =$
16) $\sqrt{289} =$
17) $\sqrt{324} =$
18) $\sqrt{400} =$
19) $\sqrt{900} =$
20) $\sqrt{529} =$
21) $\sqrt{90} =$

✎ Evaluate.

22) $8\sqrt{2} \times 2\sqrt{2} =$
23) $6\sqrt{3} - \sqrt{12} =$
24) $3\sqrt{3} + \sqrt{27} =$
25) $\sqrt{8} - \sqrt{2} =$
26) $\sqrt{27} \times \sqrt{3} =$
27) $4\sqrt{5} + \sqrt{25} =$
28) $\sqrt{169} - \sqrt{13} =$

29) $\sqrt{81} - \sqrt{3} =$
30) $\sqrt{144} + \sqrt{12} =$
31) $\sqrt{289} - \sqrt{17} =$
32) $3\sqrt{18} - 3\sqrt{2} =$
33) $\frac{3\sqrt{3}}{\sqrt{3}} =$
34) $\frac{\sqrt{4} \times \sqrt{2}}{3\sqrt{2}} =$

35) $\sqrt{36} - 5\sqrt{6} =$
36) $\sqrt{121} - 2\sqrt{11} =$
37) $\sqrt{10} \times \sqrt{6} =$
38) $\sqrt{3} \times \sqrt{5} =$
39) $\sqrt{11} \times \sqrt{3} =$
40) $\sqrt{7} + \sqrt{28} =$

✎ Solve.

41) Which of the following is equal to the square root of 75?

A. $2\sqrt{6}$

B. $36\sqrt{2}$

C. $5\sqrt{3}$

D. $12\sqrt{6}$

Answers of Worksheets

Multiplication Property of Exponents

1) $64x^{12}$
2) $2x^4$
3) $3x^5$
4) $2x^3$
5) $25x^8$
6) $4x^3y$
7) $3x^8y^2$
8) x^5y^7
9) $8x^5y^4$
10) $36x^5y^4$
11) $21x^6y^8$
12) $63x^3y^8$
13) $28x^4y^7$
14) $24x^7y^6$
15) $3x^4y^9$
16) $4x^5y^4$
17) 7^{10}
18) 6^8
19) 10^5
20) 8^{11}

Division Property of Exponents

1) $\frac{3}{2x^2}$
2) $\frac{6}{7x^3}$
3) $\frac{4x^3}{3y^8}$
4) $\frac{5y^2}{x^5}$
5) $\frac{2x^3}{7}$
6) $\frac{4y^8}{x}$
7) $\frac{4}{5x^3y^9}$
8) $\frac{6}{5x^4}$
9) $\frac{16}{9x^4y}$
10) $60a^7b^5$
11) $6b^5c^5$
12) $5b^8c^6$
13) $10a^2b^8 +$
14) $9a^7b^2c^{21}$ $35a^2b^7$
15) $4b^6c^3$

Powers of Products and Quotients

1) $8y^3x^6$
2) $16z^{12}y^{12}x^4$
3) $\frac{x^6y^3z^3}{64}$
4) $a^{20}b^6c^2$
5) $8x^6z^3y^{15}$
6) $\frac{32x^5}{z^{10}}y^{20}$
7) $8r^{15}q^6t^6$
8) $\frac{27x^6y^3z^{12}}{8m^{12}n^3}$
9) 10^{40}
10) $3^{4(a+b)}$
11) $8x^9y^{30}\frac{z^3}{27}$
12) $x^5z^{20}y^{50}$
13) $\frac{m^{30}n^{20}t^{15}}{32x^{10}}$
14) $\frac{i^{10}j^{15}k^{10}}{32}$
15) 6^{14}
16) e^{24}
17) $t^{12}\frac{p^{12}}{q^{24}}$
18) $y^{15}x^{12}\frac{1}{z^{18}}$
19) $\frac{8y^3}{z^3 \cdot x^3}$
20) $1728x^6y^6$
21) $-8j^{12}$
22) $\frac{-8x^6y^6}{z^3}$
23) $-27e^{6t(1+\omega)}$
24) 1
25) $\frac{x^6}{-8y^9z^9}$
26) $256x^8y^{16}$
27) $8x^{12}y^{12}$
28) $9x^4y^4$
29) $81x^{16}y^{12}$
30) $4x^{12}y^{16}$
31) $1,728x^{12}$
32) $8x^{45}$
33) $125x^{30}y^9$
34) $16x^{12}$
35) $225x^8$
36) $100x^{22}y^6$
37) $81x^{14}y^{10}$
38) $1,024x^{20}y^{30}$
39) $144x^2y^6$
40) $\frac{25}{x^2}$
41) x^6y^6
42) $\frac{25}{x^{10}}$
43) $\frac{x^4}{y^4}$
44) x^4y^2
45) $\frac{4y^8}{x^4}$

46) $\frac{125}{y^6}$
47) $3xyz$
48) xy
49) $3xw^6$

50) $\frac{4x^2}{(2-x)^2}$
51) $\frac{8}{27x^5y}$
52) $-2r^2q$

53) $\frac{x^8}{9z^2y^2}$
54) $8x^{12}$
55) x^2y^2
56) $4x^5y^4$

57) $3^{t+1}x^{2t+2}$
58) $\frac{2^m x^m}{3^m t^{2m}}$
59) $3^r x^r y^r z^{5r}$

Zero and Negative Exponents

1) $\frac{1}{8}$
2) $\frac{1}{64}$
3) $\frac{1}{16}$
4) $\frac{1}{100}$
5) $\frac{1}{9}$
6) $\frac{1}{9}$
7) $\frac{1}{49}$
8) $\frac{1}{81}$
9) $\frac{1}{36}$
10) $\frac{1}{125}$
11) $\frac{1}{22}$
12) $\frac{1}{16}$
13) $\frac{1}{25}$
14) $\frac{1}{35}$
15) $\frac{1}{64}$
16) $\frac{1}{216}$

17) $\frac{1}{243}$
18) $\frac{1}{25}$
19) $\frac{1}{8}$
20) $\frac{1}{27}$
21) $\frac{1}{343}$
22) $\frac{1}{216}$
23) $\frac{1}{512}$
24) $\frac{1}{81}$
25) $\frac{1}{1,000}$
26) $\frac{1}{1,000,000,000}$
27) 2
28) 4
29) 9
30) $\frac{9}{4}$
31) 125
32) $\frac{16}{9}$

33) $\frac{25}{4}$
34) 256
35) $\frac{125}{8}$
36) $\frac{49}{9}$
37) $\frac{216}{125}$
38) $\frac{1}{e^{4t}x^2}$
39) $\frac{9x^2}{z^6}$
40) $a^{12}b^2$
41) $\frac{64z^6}{x^6y^3}$
42) $\frac{8k^3}{g^3h^9}$
43) $\frac{x}{sqr^2}$
44) 1
45) $\frac{x^3}{2y}$
46) $\frac{1}{x^6z^9}$
47) $\frac{9}{x^2y^4}$

48) $4x^2$
49) $\frac{1}{9x^2y^2}$
50) 1
51) $\frac{1}{(3x+y)^2}$
52) $\frac{729x^3}{4x+1}$
53) $\frac{z^3x^3}{8y^3}$
54) 32
55) $\frac{(2-2x^2)^2}{9}$
56) 1
57) $\frac{1}{(1-x)^2}$
58) $\frac{1}{x^4}$
59) $\frac{x^4}{z^2y^2}$
60) $\frac{z^2}{x^2}$
61) $\frac{z^2}{y}$
62) $\frac{1}{8x}$

Writing Scientific Notation

1) 1.13×10^{-1}
2) 2×10^{-2}
3) 2.5×10^0

4) 2×10^1
5) 6×10^1
6) 4×10^{-3}

7) 7.8×10^1
8) 1.6×10^3
9) 1.45×10^3

10) 9.1×10^4
11) 2×10^6
12) 6×10^{-7}
13) 3.54×10^5
14) 3.25×10^{-4}

15) 2.3×10^{-4}
16) 5.6×10^7
17) 2.1×10^4
18) 7.8×10^7
19) 2.2×10^{-6}

20) 1.2×10^{-4}
21) 2×10^{-2}
22) 8.8×10^6
23) 1.21×10^7
24) 1.8×10^{-1}

Square Roots

1) 1
2) 2
3) 3
4) 5
5) 4
6) 7
7) 6
8) 0
9) 8
10) 9
11) 11
12) 15

13) 12
14) 10
15) 16
16) 17
17) 18
18) 20
19) 30
20) 23
21) $3\sqrt{10}$
22) 32
23) $4\sqrt{3}$
24) $6\sqrt{3}$

25) $\sqrt{2}$
26) 9
27) $4\sqrt{5} + 5$
28) $13 - \sqrt{13}$
29) $9 - \sqrt{3}$
30) $12 + 2\sqrt{3}$
31) $17 - \sqrt{17}$
32) $6\sqrt{2}$
33) 3
34) $\frac{2}{3}$

Section 3:

Geometry and Statistics

- *Plane Figures*

- *Solid Figures*

- *Statistics*

- *Probability*

Chapter 11:

Plane Figures

Topics that you'll learn in this part:

- ✓ Transformations: Translations, Rotations, and Reflections
- ✓ The Pythagorean Theorem
- ✓ Area of Triangles
- ✓ Perimeter of Polygons
- ✓ Area and Circumference of Circles
- ✓ Area of Squares, Rectangles, and Parallelograms
- ✓ Area of Trapezoids

Transformations: Translations, Rotations, and Reflections

✎ *Graph the image of the figure using the transformation given.*

1) translation: 4 units right and 1 unit down

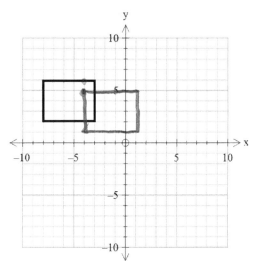

2) translation: 4 units left and 2 unit up

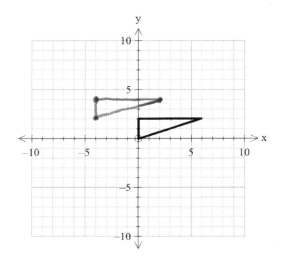

3) rotation 90° counterclockwise about the origin

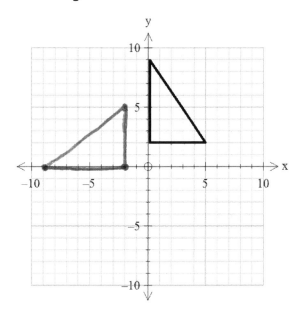

4) rotation 180° about the origin

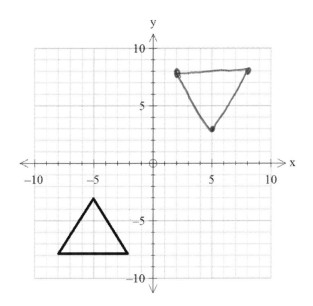

The Pythagorean Theorem

✎ *Do the following lengths form a right triangle?*

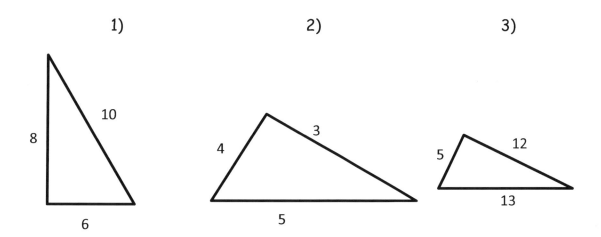

✎ *Find each missing length to the nearest tenth.*

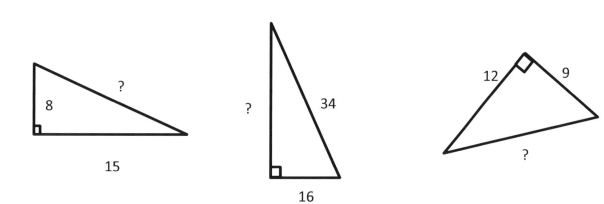

Area of Triangles

🖉 *Find the area of each.*

1)
$c = 12\ mi$
$h = 4\ mi$
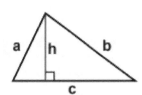

2)
$s = 15\ m$
$h = 9m$
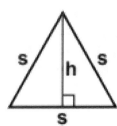

3)
$a = 5\ m$
$b = 11\ m$
$c = 14\ m$
$h = 4\ m$
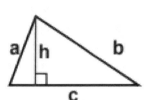

4)
$s = 10\ m$
$h = 8.6\ m$

5)
$c = 15\ mi$
$h = 6\ mi$
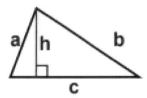

6)
$a = 5\ m$
$h = 4\ m$
$b = 9$
$C = 12$
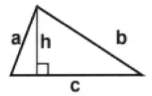

7)
$c = 8\ mi$
$h = 4\ mi$
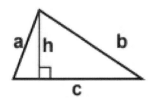

8)
$s = 10\ m$
$h = 8\ m$
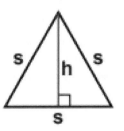

Perimeter of Polygon

📝 *Find the perimeter of each shape.*

1)
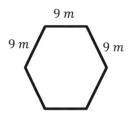

2)
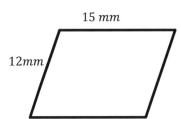

3)
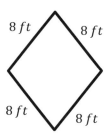

4)
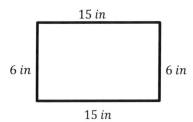

5)
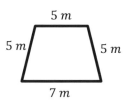

6)
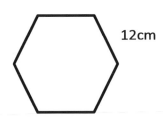

7)
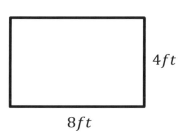

8)
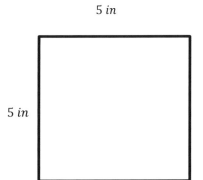

Area and Circumference of Circles

✏️ *Find the area and circumference of each.* (π = 3.14)

1)

2)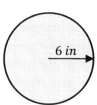

3)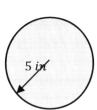

4)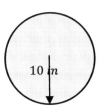

✏️ *Find the area and of each.* (π = 3.14)

5)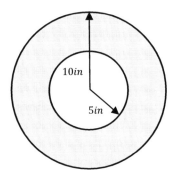

6)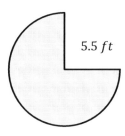

7)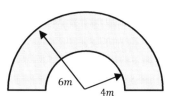

8)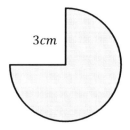

Area of Squares, Rectangles, and Parallelograms

🖉 Find the area of each.

1)
12 yd
21yd   21yd
12yd

2)
17mi
17 mi   17 mi
17 mi

3)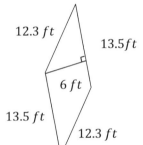

4)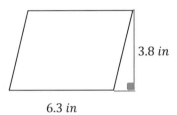

5)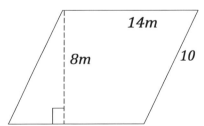

6)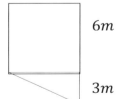

Area of Trapezoids

✏️ *Calculate the area for each trapezoid.*

1)
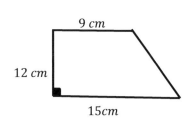

2)
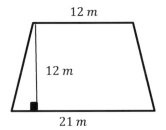

3)
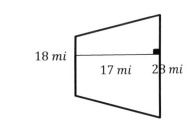

4)
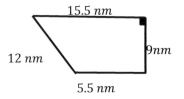

5)
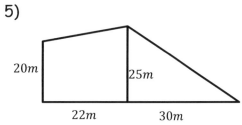

6)
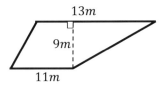

7)
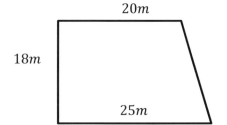

8)
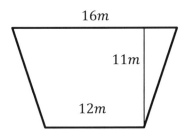

Answers of Worksheets

Transformations

1) translation:

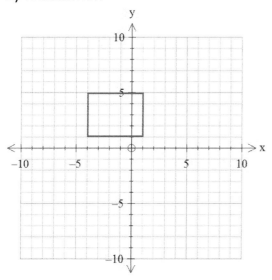

2) translation:

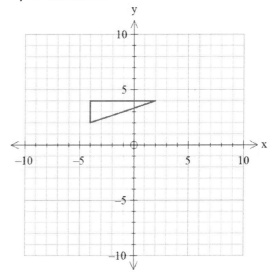

3) rotation 90° counterclockwise about the origin

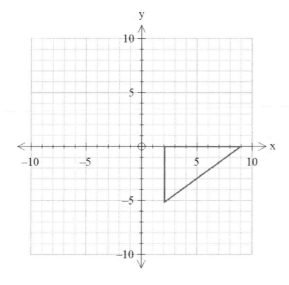

4) rotation 180° about the origin

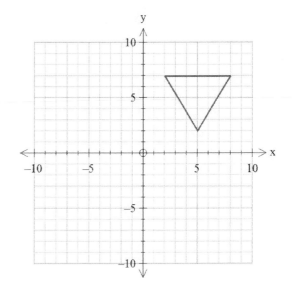

The Pythagorean Theorem

1) yes

2) yes

3) yes

4) 17

5) 30

6) 15

Area of Triangles

1) $24 mi^2$
2) $67.5 m2$
3) $28 m2$
4) $43 m2$
5) $45 m^2$
6) $24 m^2$
7) $16 mi^2$
8) $40 m^2$

Perimeter of Polygons

1) $54 m$
2) $54 mm$
3) $32 ft$
4) $42 in$
5) $22 m$
6) $72 cm$
7) $24 ft$
8) $20 in$

Area and Circumference of Circles

1) Area: $50.26 in^2$, Circumference: $25.12 in$
2) Area: $113.1 in^2$, Circumference: $31.7 in$
3) Area: $78.5 in^2$, Circumference: $31.4 in$
4) Area: $314.16 in^2$, Circumference: $62.83 in$
5) Area: $235.62 in^2$
6) Area: $71.27 ft^2$
7) Area: $31.415 m^2$
8) Area: $21.2 cm^2$

Area of Squares, Rectangles, and Parallelograms

1) $252 yd^2$
2) $289 mi^2$
3) $81 ft^2$
4) $23.94 in^2$
5) $112 m^2$
6) $45 m^2$

Area of Trapezoids

1) $144 cm^2$
2) $198 m^2$
3) $391 mi^2$
4) $94.5 nm^2$
5) $870 m^2$
6) $108 m^2$
7) $405 m^2$
8) $154 m^2$

Chapter 12:

Solid Figures

Topics that you'll learn in this part:

- ✓ Volume of Cubes and Rectangle Prisms
- ✓ Surface Area of Cubes
- ✓ Surface Area of a Prism
- ✓ Volume of Pyramids and Cones
- ✓ Surface Area of Pyramids and Cones

Volume of Cubes and Rectangle Prisms

✎ *Find the volume of each of the rectangular prisms.*

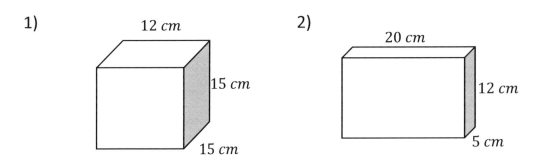

1) 12 cm, 15 cm, 15 cm

2) 20 cm, 12 cm, 5 cm

✎ *Solve.*

3) Layla wants to build a wooden box with a volume of 45 cubic centimeters. She started with a width of 3cm. How long should Layla make the box?

4) The sea turtle habitat at the zoo is made by connecting two large aquariums. The first aquarium is 6m long, 4m wide, and 2m high. The second aquarium is 8m long, 9m wide, and 3m high. How many cubic meters of space do the sea turtles have in their habitat?

5) The closet is 6 feet wide, 5 feet deep and 8 feet tall. In the closet, there is a suitcase that is 2 feet wide, 3 feet long and 4 feet tall. How much room is left in the closet?

6) Find the volume of the rectangular prism.

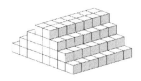

Surface Area of Cubes

✏️ *Find the surface of each cube.*

1)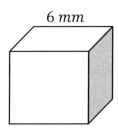
6 mm

2)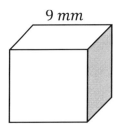
9 mm

3)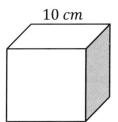
10 cm

4)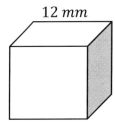
12 mm

5)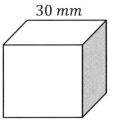
30 mm

6)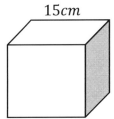
15 cm

7)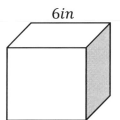
6 in

8)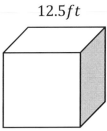
12.5 ft

9)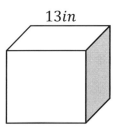
13 in

10)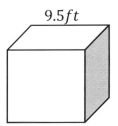
9.5 ft

Surface Area of a Prism

✎ Find the surface of each prism.

1)
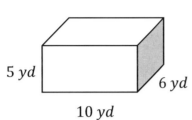

2)
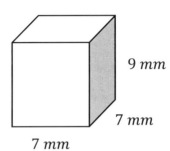

3)
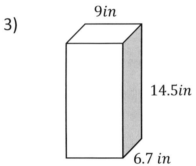

4)
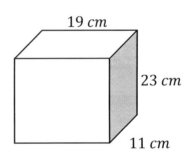

5)
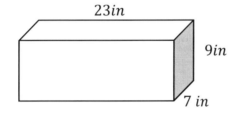

6)
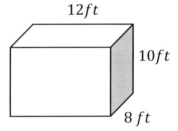

7)
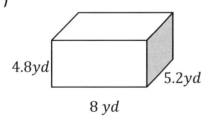

8)
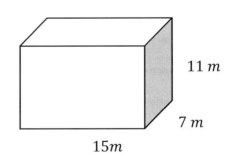

Volume of Pyramids and Cones

✎ *Find the volume of each figure.* (π = 3.14)

1)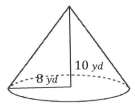

2)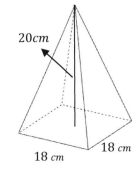

3)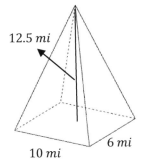

4)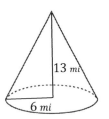

5)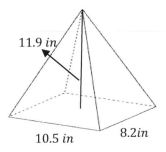

6)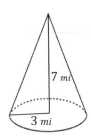

7)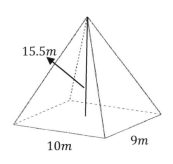

8)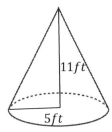

Answers of Worksheets

Volume of Cubes and Rectangle Prisms

1) $2,700 cm^3$
2) $1,200 cm^3$
3) 5
4) 264
5) 216
6) 140

Surface Area of a Cube

1) $216\ mm^2$
2) $486\ mm^2$
3) $600\ cm^2$
4) $864\ mm^2$
5) $5,400\ mm^2$
6) $1,350\ cm^2$
7) $216\ in^2$
8) $937.5\ ft^2$
9) $1,014\ in^2$
10) $541.5\ ft^2$

Surface Area of a Prism

1) $240\ yd^2$
2) $350\ mm^2$
3) $576.6\ in^2$
4) $1,798\ cm^2$
5) $736\ in^2$
6) $592\ ft^2$
7) $209.92\ yd^2$
8) $694\ m^2$

Volume of Pyramids and Cones

1) $670.2\ yd^3$
2) $2,160 cm^3$
3) $250 mi^3$
4) $490.1 mi^3$
5) $341.5 in^3$
6) $65.97 mi^3$
7) $465 m^3$
8) $288 ft^3$

Chapter 13:

Statistics

Topics that you'll learn in this part:

✓ Mean, Median, Mode, and Range of the Given Data
✓ First Quartile, Second Quartile and Third Quartile of the Given Data
✓ Bar Graph
✓ Box and Whisker Plots
✓ Stem–And–Leaf Plot
✓ The Pie Graph or Circle Graph
✓ Scatter Plots

Mean, Median, Mode, and Range of the Given Data

✏️ *Find Mean, Median, Mode, and Range of the Given Data.*

1) 7, 2, 5, 1, 1, 2, 3, 4
2) 2, 2, 2, 3, 6, 3, 7, 4
3) 9, 4, 3, 1, 7, 9, 4, 6, 4
4) 8, 4, 2, 4, 3, 2, 4, 5
5) 8, 5, 7, 5, 7, 9, 8, 8, 6
6) 5, 1, 4, 4, 9, 2, 9, 1, 2, 5, 1, 8
7) 4, 7, 5, 9, 5, 7, 7, 7, 5, 2, 3, 5
8) 7, 5, 4, 9, 6, 7, 7, 5, 2, 8
9) 2, 5, 5, 6, 2, 4, 7, 6, 4, 9, 5
10) 10, 5, 2, 5, 4, 5, 8, 10, 8

11) 4, 5, 2, 2, 6, 8, 10, 12
12) 5, 9, 5, 9, 8, 6, 11, 9, 6, 8
13) 14, 16, 16, 15, 19, 16
14) 10, 9, 12, 13, 13, 17, 15, 10
15) 3, 2, 9, 8, 5, 5, 6, 8
16) 14, 16, 18, 17, 12, 16, 15, 16
17) 9, 18, 17, 15, 14, 19, 18, 17
18) 15, 12, 18, 17, 15, 15, 12, 14
19) 32, 51, 38, 69, 15, 50, 38, 8
20) 1, 9, 8, 6, 5, 9, 8, 9

✏️ *Solve.*

21) A stationery sold 14 pencils, 40 red pens, 50 blue pens, 10 notebooks, 16 erasers, 38 rulers and 36 color pencils. What are the Mode and Range for the stationery sells?

22) In an English test, eight students score 14, 13, 17, 11, 19, 20, 14 and 15. What are their Median, Mode and Range?

23) Bob has 12 black pen, 14 red pen, 15 green pens, 24 blue pens and 3 boxes of yellow pens. If the Mean and Median are 16 and 15 respectively, what is the number of yellow pens in each box?

Box and Whisker Plot

✎ *Make box and whisker plots for the given data.*

1) 11,17,22,18,23,2,3,16,21,7,8,15,5

2) 33,31,30,38,40,36

3) 46,36,15,21,65,25,48,70,68

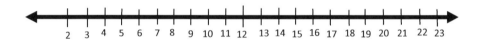

4) 9,10,12,15,17,19,24,26,28

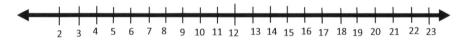

5) 41,43,45,47,51,50,44

6) 60,66,62,65,68,70,72

7) 72,82,81,76,77,82,84,79,80

Bar Graph

🖋 *Graph the given information as a bar graph.*

1) The number of bed-sheets manufactured by a factory during five consecutive weeks is given below. Draw the bar graph representing the above data.

week	First	Second	Third	Fourth	Fifth
Number of bed-sheets	550	810	680	320	850

2) The number of students in 7 different classes is given below. Represent this data on the bar graph.

class	6th	7th	8th	9th	10th	11th	12th
Number of students	125	115	130	140	145	100	80

3) The number of trees planted by Eco-club of a school in different years is given below. Draw the bar graph to represent the data.

Year	2005	2006	2007	2008	2009	2010
Number of trees to be planted	150	220	350	400	300	380

4) The following data represents the sale of refrigerator sets in a showroom in first 6 months of the year. Draw the bar graph for the data given and find out the months in which the sale was minimum and maximum.

Months	Jan	Feb	March	April	May	June
No. of refrigerator sold	19	21	12	46	35	28

Stem–And–Leaf Plot

Make stem ad leaf plots for the given data.

1) 74,88,97,72,79,86,95,79,83,91

   Stem | Leaf plot

2) 37,48,26,33,49,26,19,26,48

   Stem | Leaf plot

3) A zookeeper created the following stem-and-leaf plot showing the number of tigers at each major zoo in the country. What was the smallest number of tigers at any one zoo?

Stem	Leaf plot
0	7
1	1 4 8
2	5 5 5 6 7 7 9
3	
4	

4) The government published the following stem-and-leaf plot showing the number of bears at each major zoo in the country. How many zoos have more than 50 bears?

Stem	Leaf plot
0	
1	8 8
2	1 3 6 6 7 7 9
3	1 7
4	1 4 5 7
5	0 3

The Pie Graph or Circle Graph

✏ Solve.

1) Suppose you take a poll of the students in your class to find out their favorite foods, and get the following results:

Pizza: 41%, Ice Cream: 24%, Raw Mushrooms: 9%, Dog Food: 11%, Chicken Livers: 15%

Organize this data in a circle graph.

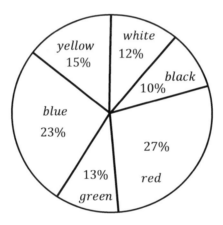

Favorite colors

2) Which color is the most?

3) What percentage of pie graph is yellow?

4) Which color is the least?

5) What percentage of pie graph is blue?

6) What percentage of pie graph is green?

Scatter Plots

✎ Construct a scatter plot.

1) Construct a graph of the length of the humerus bone vs. the length of the radius.

Length of Radius (cm)	25	22	23.5	22.5	23	22.6	21.4	21.9	23.5	24.3	24
Length of Humerus (cm)	29.7	26.5	27.1	26	28	25.2	24	23.8	26.7	29	27

2) Plot the data on the scatter plot below, choosing appropriate scales and labels.

Age	25	30	35	37	38	40	41	45	55	60	62	65	70	75
Earnings ($)	22000	26500	29500	29000	30000	32000	35000	36000	41000	41000	42500	43000	37000	37500

3) The table shows the numbers of students remaining on an after-school bus and the numbers of minutes since leaving the school.

Minutes	0	5	9	15	23	26	32
Number of students	56	45	39	24	17	6	0

Plot the data from the table on the graph. Describe the relationship between the two data sets.

Answers of Worksheets

Mean, Median, Mode, and Range of the Given Data

1) mean: 3.125, median: 2.5, mode: 1, 2, range: 6

2) mean: 3.625, median: 3, mode: 2, range: 5

3) mean: 5.22, median: 4, mode: 4, range: 8

4) mean: 4, median: 4, mode: 4, range: 6

5) mean: 7, median: 7, mode: 5, 7, 8, range: 4

6) mean: 4.25, median: 4, mode: 1, range: 8

7) mean: 5.5, median: 5, mode: 7.5, range: 7

8) mean: 6, median: 6.5, mode: 7, range: 7

9) mean: 5, median: 5, mode: 5, range: 7

10) mean: 6.33, median: 5, mode: 5, range: 8

11) mean: 6.125, median: 2, mode: 2, range: 10

12) mean: 7.6, median:8, mode: 9, range: 6

13) mean: 16, median:16, mode: 16, range: 5

14) mean: 12.375, median:12.5, mode: 10,13, range: 8

15) mean: 5.75, median:5.5, mode:8,5, range: 7

16) mean: 15.5, median:16, mode: 16, range: 6

17) mean: 15.875, median:17, mode: 18,17, range: 10

18) mean: 14.75, median:15, mode: 15, range: 6

19) mean: 37.625, median:38, mode: 38, range: 61

20) mean: 6.875, median:8, mode: 9, range: 8

21) Mode: none, range:40

22) median:14.5, mode:14, range:9

23) 5

Box and Whisker Plots

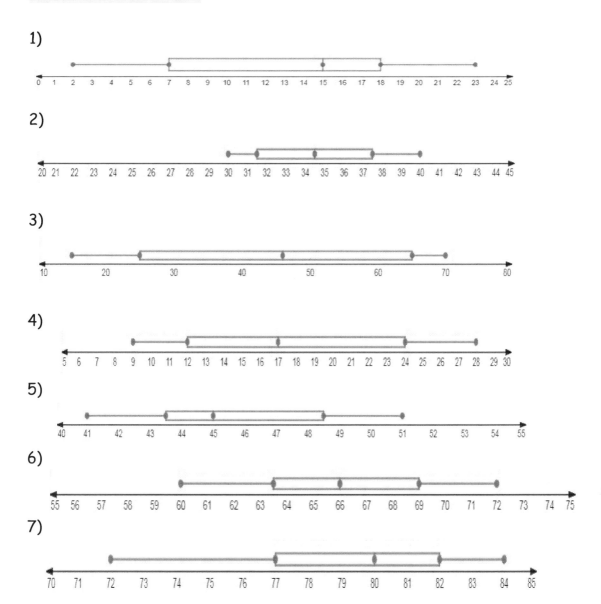

Bar Graph

1)

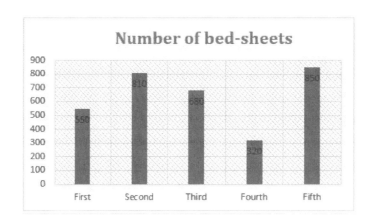

2)

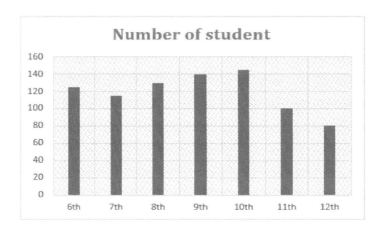

3)

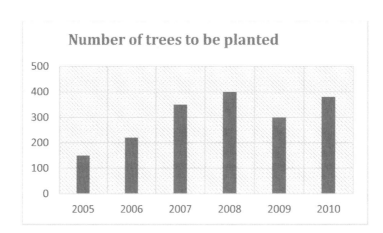

4)

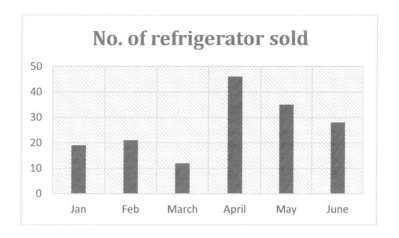

Stem–And–Leaf Plot

1)

Stem	leaf
7	2 4 9 9
8	3 6 8
9	1 5 7

2)

Stem	leaf
1	9
2	6 6 6
3	3 7
4	8 8 9

3)

Stem	leaf
4	1 2
5	3 4 4 8
6	5 5 7 9

4) 7
5) 1

The Pie Graph or Circle Graph

1)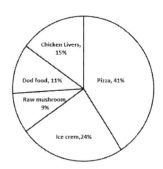

2) red

3) 15%

4) black

5) 23%

6) 13%

Scatter Plots

1)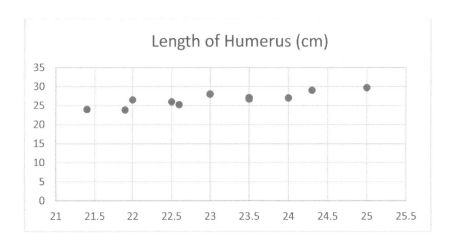

2)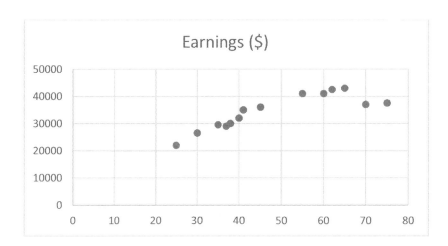

3)

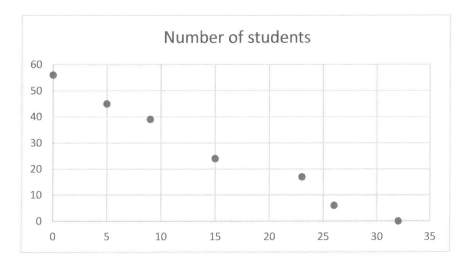

When time increase, number of students decrease

Chapter 14:

Probability

Topics that you'll learn in this part:

✓ Probability of Simple Events

✓ Experimental Probability

✓ Independent and Dependent Events Word Problems

✓ Factorials

✓ Permutations

✓ Combination

Probability of Simple Events

Solve.

1) A number is chosen at random from 1 to 50. Find the probability of selecting multiples of 10.

2) You spin the spinner shown below once. Each sector shown has an equal area. What is P (shaded sector)?

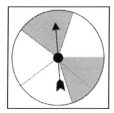

3) You draw a card at random from a deck that contains 3 black cards and 7 red cards. What is the probability of choosing a black card?

4) Greg has an MP3 player called the Jumble. The Jumble randomly selects a song for the user to listen to. Greg's Jumble has 6 classical songs, 7 rock songs, and 9 rap songs on it. What is P (not a rap song)?

5) Omar ordered his sister a birthday card from a company that randomly selects a card from their inventory. The company has 21 total cards in inventory. 14 of those cards are birthday cards. What is P (not a birthday card)?

6) Giovanna owns a farm. She is going to randomly select one animal to present at the state fair. She has 6 pigs, 7 chickens, and 10 cows. What is What is the probability of choosing a chicken?

Experimental Probability

✎ Solve.

1) A bag contains 18 balls: two green, five black, eight blue, a brown, a red and one white. If 17 balls are removed from the bag at random, what is the probability that a brown ball has been removed?

   A. $\frac{1}{9}$

   B. $\frac{1}{6}$

   C. $\frac{16}{17}$

   D. $\frac{17}{18}$

2) The table shows the number of feathers Patsy the Peacock sold at each of the 8 festivals this year. Based on this data, what is a reasonable estimate of the probability that Patsy sells fewer than 5 feathers next festival?

3	6	1	4
2	3	7	2

3) The Cinemania theater showed 108 different movies last year. Of those, 15 movies were action movies. Based on this data, what is a reasonable estimate of the probability that the next movie is an action movie?

4) Stephen read 12 books, 20 magazines, and 17 newspaper articles last year. Based on this data, what is a reasonable estimate of the probability that Stephen's next reading material is a magazine?

5) March Madness Movies served 23 lemonades out of a total of 111 fountain drinks last weekend. Based on this data, what is a reasonable estimate of the probability that the next fountain drink ordered is a lemonade?

Factorials

📝 *Determine the value for each expression*

1) $\frac{9!}{6!}$

2) $\frac{10!}{8!}$

3) $\frac{7!}{3!}$

4) $\frac{5!}{4!}$

5) $\frac{6!+2!}{4!}$

6) $3! \times 4!$

📝 *Solve.*

7) While playing Scrabble®, you need to make a word out of the letters A N P S. How many arrangements of these letters are possible?

8) How many ways can you have your quarter pound hamburger prepared if you can have it prepared with or without mustard, ketchup, mayonnaise, lettuce, tomatoes, pickles, cheese, and onions?

9) How many arrangements of all the letters in the word PYRAMID do not end with D?

10) Five students are running for Junior Class President. They must give speeches before the election committee. In how many different orders could they give their speeches?

11) How many five-letter "words" can be formed from the letters in the word COMBINE?

12) In how many ways can six different algebra books and three different geometry books be arranged on a shelf if all the books of one subject must remain together?

Combination and Permutations

✏️ *Solve.*

1) In how many ways can we select a chairman, vice-chairman, secretary, and treasurer from a group of 10 persons?

2) You have been asked to judge an art contest with 15 entries. In how many ways can you assign 1st, 2nd and 3rd place? (Express your answer as P (n, k) for some n and k and evaluate.)

3) How many three letter words (including nonsense words) can you make from the letters of the English alphabet, if letters cannot be repeated? (Express your answer as P (n, k) for some n and k and evaluate.)

4) Five students are to be chosen from a class of 10 and lined up for a photograph. How many such photographs can be taken?

5) You have 6 reindeer, Prancer, Rudy, Balthazar, Quentin, Jebediah, and Lancer, and you want to have 3 fly your sleigh. You always have your reindeer fly in a single-file line. How many ways can you arrange your reindeer?

6) A committee of 5 people is to be chosen from a group of 6 men and 4 women. How many committees are possible if there are no restrictions?

7) In a hand of poker, 5 cards are dealt from a regular pack of 52 cards. What is the total possible number of hands if there are no restrictions?

8) You just got a free ticket for a boat ride, and you can bring along 2 friends! Unfortunately, you have 5 friends who want to come along. How many different groups of friends could you take with you?

9) Christopher is packing his bags for his vacation. He has 8 unique shirts, but only 5 fit in his bag. How many different groups of 5 shirts can he take?

Answers of Worksheets

Probability of simple events

1) $\frac{5}{10}$ 2) $\frac{2}{5}$ 3) $\frac{3}{10}$ 4) $\frac{13}{22}$ 5) $\frac{7}{21}$ 6) $\frac{1}{23}$

Experimental Probability

1) $\frac{17}{18}$ 2) 0.75 3) $\frac{15}{108}$ 4) $\frac{20}{49}$ 5) $\frac{23}{111}$

Factorials

1) 504
2) 90
3) 840
4) 5
5) 30.08
6) 144
7) 24
8) 256
9) 4320
10) 120
11) 2520
12) 720

Combination and Permutations

1) 5040
2) 2730
3) 15600
4) 30240
5) 120
6) 252
7) 2598960
8) 10
9) 56

Time to Test

Time to refine your skill with a practice examination

Take practice ISEE Upper Level Math Tests to simulate the test day experience. After you've finished, score your test using the answer key.

Before You Start

- You'll need a pencil and scratch papers to take the test.
- It's okay to guess. You won't lose any points if you're wrong.
- After you've finished the test, review the answer key to see where you went wrong.

Calculators are NOT permitted on the ISEE Upper Level Test.

Good Luck!

ISEE Upper Level Practice Tests Answer Sheets

Remove (or photocopy) this answer sheet and use it to complete the practice test.

ISEE Upper Level Mathematics Achievement Test 1

1) Ⓐ Ⓑ Ⓒ Ⓓ   2) Ⓐ Ⓑ Ⓒ Ⓓ
3) Ⓐ Ⓑ Ⓒ Ⓓ   4) Ⓐ Ⓑ Ⓒ Ⓓ
5) Ⓐ Ⓑ Ⓒ Ⓓ   6) Ⓐ Ⓑ Ⓒ Ⓓ
7) Ⓐ Ⓑ Ⓒ Ⓓ   8) Ⓐ Ⓑ Ⓒ Ⓓ
9) Ⓐ Ⓑ Ⓒ Ⓓ   10) Ⓐ Ⓑ Ⓒ Ⓓ
11) Ⓐ Ⓑ Ⓒ Ⓓ   12) Ⓐ Ⓑ Ⓒ Ⓓ
13) Ⓐ Ⓑ Ⓒ Ⓓ   14) Ⓐ Ⓑ Ⓒ Ⓓ
15) Ⓐ Ⓑ Ⓒ Ⓓ   16) Ⓐ Ⓑ Ⓒ Ⓓ
17) Ⓐ Ⓑ Ⓒ Ⓓ   18) Ⓐ Ⓑ Ⓒ Ⓓ
19) Ⓐ Ⓑ Ⓒ Ⓓ   20) Ⓐ Ⓑ Ⓒ Ⓓ
21) Ⓐ Ⓑ Ⓒ Ⓓ   22) Ⓐ Ⓑ Ⓒ Ⓓ
23) Ⓐ Ⓑ Ⓒ Ⓓ   24) Ⓐ Ⓑ Ⓒ Ⓓ
25) Ⓐ Ⓑ Ⓒ Ⓓ   26) Ⓐ Ⓑ Ⓒ Ⓓ
27) Ⓐ Ⓑ Ⓒ Ⓓ   28) Ⓐ Ⓑ Ⓒ Ⓓ
29) Ⓐ Ⓑ Ⓒ Ⓓ   30) Ⓐ Ⓑ Ⓒ Ⓓ
31) Ⓐ Ⓑ Ⓒ Ⓓ   32) Ⓐ Ⓑ Ⓒ Ⓓ

33) Ⓐ Ⓑ Ⓒ Ⓓ   34) Ⓐ Ⓑ Ⓒ Ⓓ
35) Ⓐ Ⓑ Ⓒ Ⓓ   36) Ⓐ Ⓑ Ⓒ Ⓓ
37) Ⓐ Ⓑ Ⓒ Ⓓ   38) Ⓐ Ⓑ Ⓒ Ⓓ
39) Ⓐ Ⓑ Ⓒ Ⓓ   40) Ⓐ Ⓑ Ⓒ Ⓓ
41) Ⓐ Ⓑ Ⓒ Ⓓ   42) Ⓐ Ⓑ Ⓒ Ⓓ
43) Ⓐ Ⓑ Ⓒ Ⓓ   44) Ⓐ Ⓑ Ⓒ Ⓓ
45) Ⓐ Ⓑ Ⓒ Ⓓ   46) Ⓐ Ⓑ Ⓒ Ⓓ
47) Ⓐ Ⓑ Ⓒ Ⓓ

ISEE Upper Level Mathematics Achievement Test 2

1) Ⓐ Ⓑ Ⓒ Ⓓ   2) Ⓐ Ⓑ Ⓒ Ⓓ
3) Ⓐ Ⓑ Ⓒ Ⓓ   4) Ⓐ Ⓑ Ⓒ Ⓓ
5) Ⓐ Ⓑ Ⓒ Ⓓ   6) Ⓐ Ⓑ Ⓒ Ⓓ
7) Ⓐ Ⓑ Ⓒ Ⓓ   8) Ⓐ Ⓑ Ⓒ Ⓓ
9) Ⓐ Ⓑ Ⓒ Ⓓ   10) Ⓐ Ⓑ Ⓒ Ⓓ
11) Ⓐ Ⓑ Ⓒ Ⓓ   12) Ⓐ Ⓑ Ⓒ Ⓓ
13) Ⓐ Ⓑ Ⓒ Ⓓ   14) Ⓐ Ⓑ Ⓒ Ⓓ
15) Ⓐ Ⓑ Ⓒ Ⓓ   16) Ⓐ Ⓑ Ⓒ Ⓓ
17) Ⓐ Ⓑ Ⓒ Ⓓ   18) Ⓐ Ⓑ Ⓒ Ⓓ
19) Ⓐ Ⓑ Ⓒ Ⓓ   20) Ⓐ Ⓑ Ⓒ Ⓓ

21) Ⓐ Ⓑ Ⓒ Ⓓ	22) Ⓐ Ⓑ Ⓒ Ⓓ
23) Ⓐ Ⓑ Ⓒ Ⓓ	24) Ⓐ Ⓑ Ⓒ Ⓓ
25) Ⓐ Ⓑ Ⓒ Ⓓ	26) Ⓐ Ⓑ Ⓒ Ⓓ
27) Ⓐ Ⓑ Ⓒ Ⓓ	28) Ⓐ Ⓑ Ⓒ Ⓓ
29) Ⓐ Ⓑ Ⓒ Ⓓ	30) Ⓐ Ⓑ Ⓒ Ⓓ
31) Ⓐ Ⓑ Ⓒ Ⓓ	32) Ⓐ Ⓑ Ⓒ Ⓓ
33) Ⓐ Ⓑ Ⓒ Ⓓ	34) Ⓐ Ⓑ Ⓒ Ⓓ
35) Ⓐ Ⓑ Ⓒ Ⓓ	36) Ⓐ Ⓑ Ⓒ Ⓓ
37) Ⓐ Ⓑ Ⓒ Ⓓ	38) Ⓐ Ⓑ Ⓒ Ⓓ
39) Ⓐ Ⓑ Ⓒ Ⓓ	40) Ⓐ Ⓑ Ⓒ Ⓓ
41) Ⓐ Ⓑ Ⓒ Ⓓ	42) Ⓐ Ⓑ Ⓒ Ⓓ
43) Ⓐ Ⓑ Ⓒ Ⓓ	44) Ⓐ Ⓑ Ⓒ Ⓓ
45) Ⓐ Ⓑ Ⓒ Ⓓ	46) Ⓐ Ⓑ Ⓒ Ⓓ
47) Ⓐ Ⓑ Ⓒ Ⓓ	

ISEE Upper Level

Mathematics Practice Test 1

- 47 questions
- **Total time for this section:** 40 Minutes
- **Calculators are not allowed at the test.**

1) Solve.

$$|15 - (18 \div |2 - 8|)| = ?$$

A. 8

B. -12

C. 12

D. -8

2) $(x - 4)(2x + 2) =$

A. $2x^2 + 2x + 8$

B. $2x^2 - 4x - 8$

C. $2x^2 + 8x + 12$

D. $2x^2 - 6x - 8$

3) Which of the following graphs represents the compound inequality $-4 \leq 4x - 5 < 7$?

A.

B.

C.

D.

4) Find all values of x for which $3x^2 + 8x + 4 = 0$

A. $-\frac{3}{2}, -\frac{1}{2}$

B. $-\frac{1}{2}, -3$

C. $-2, -\frac{1}{3}$

D. $-\frac{2}{3}, -2$

5) Which value of x makes the following inequality true?

$$\frac{8}{23} \leq x < 45\%$$

A. 0.31

B. $\frac{8}{19}$

C. $\sqrt{0.052}$

D. 0.512

6) Which graph shows a non-proportional linear relationship between x and y?

A.
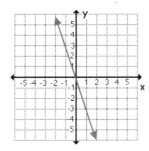

B.
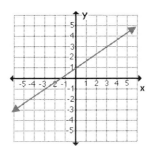

C.
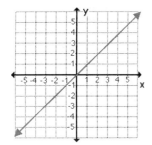

D.
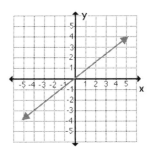

7) The rectangle on the coordinate grid is translated 6 units down and 5 units to the left.

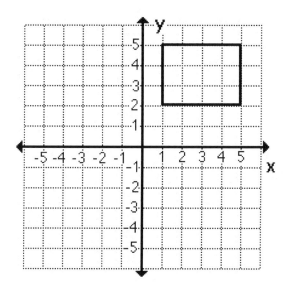

Which of the following describes this transformation?

A. $(x, y) \Rightarrow (x - 6, y - 5)$

B. $(x, y) \Rightarrow (x - 6, y + 5)$

C. $(x, y) \Rightarrow (x + 5, y + 6)$

D. $(x, y) \Rightarrow (x - 5, y - 6)$

8) A girl 150 cm tall, stands 280 cm from a lamp post at night. Her shadow from the light is 100 cm long. How high is the lamp post?

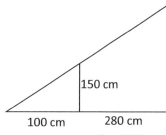

A. 550

B. 570

C. 590

D. 600

9) How is this number written in scientific notation?

0.00000625

A. 6.25×10^{-8}

B. 6.25×10^{-6}

C. 0.625×10^{-7}

D. 62.5×10^{-8}

10) The ratio of boys to girls in a school is 7:5. If there are 540 students in a school, how many boys are in the school?

A. 285

B. 315

C. 345

D. 385

11) There are 5 blue marbles, 7 red marbles, and 6 yellow marbles in a box. If Ava randomly selects a marble from the box, what is the probability of selecting a red or yellow marble?

A. $\frac{7}{13}$

B. $\frac{1}{3}$

C. $\frac{13}{18}$

D. $\frac{18}{13}$

12) Emily and Daniel have taken the same number of photos on their school trip. Emily has taken 9 times as many as photos as Claire and Daniel has taken 24 more photos than Claire. How many photos has Claire taken?

A. 3

B. 4

C. 6

D. 10

13) Emily lives $4\frac{2}{5}$ miles from where she works. When traveling to work, she walks to a bus stop $\frac{1}{4}$ of the way to catch a bus. How many miles away from her house is the bus stop?

A. $1\frac{1}{3}$ Miles

B. $1\frac{1}{10}$ Miles

C. $1\frac{3}{10}$ Miles

D. $1\frac{3}{5}$ Miles

14) Use the diagram below to answer the question.

Given the lengths of the base and diagonal of the rectangle below, what is the length of height h, in terms of s?

A. $8s\sqrt{2}$

B. $4s\sqrt{7}$

C. $5s$

D. $8s$

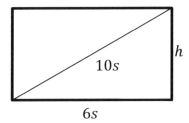

15) $5 - 12 \div (2^4 \div 4) =$ ___

   A. -2

   B. $\frac{3}{4}$

   C. 2

   D. 3

16) If the area of trapezoid is 36 cm, what is the perimeter of the trapezoid?

   A. 12 cm

   B. 22 cm

   C. 32 cm

   D. 42 cm

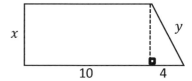

17) If a vehicle is driven 40 miles on Monday, 30 miles on Tuesday, and 38 miles on Wednesday, what is the average number of miles driven each day?

   A. 29 Miles

   B. 31 Miles

   C. 32 Miles

   D. 36 Miles

18) Find the area of a rectangle with a length of 212 feet and a width of 52 feet.

   A. 12,296sq. ft

   B. 12,521sq. ft

   C. 12,396sq. ft

   D. 12,855sq. ft

19) If $(6.2+6.3+6.5)x = x+2$, then what is the value of x?

   A. 0

   B. $\frac{1}{9}$

   C. 1

   D. 10

20) With an 16% discount, Ella was able to save $15.4 on a dress. What was the original price of the dress?

   A. $94.56

   B. $95.62

   C. $96.25

   D. $98.52

21) $\frac{9}{30}$ is equals to:

   A. 0.03

   B. 0.27

   C. 0.29

   D. 0.3

22) The sum of 8 numbers is greater than 240 and less than 320. Which of the following could be the average (arithmetic mean) of the numbers?

A. 30

B. 35

C. 40

D. 45

23) $75 \div \frac{1}{6} = ?$

A. 12.5

B. 80

C. 230

D. 450

24) Two dice are thrown simultaneously, what is the probability of getting a sum of 5 or 7?

A. $\frac{1}{3}$

B. $\frac{1}{4}$

C. $\frac{1}{6}$

D. $\frac{5}{18}$

25) Simplify $\dfrac{\frac{1}{4} - \frac{x+7}{8}}{\frac{x^2}{4} - \frac{1}{4}}$

A. $\dfrac{3 - x}{x^2 - 2}$

B. $\dfrac{-9 - x}{2x^2 - 2}$

C. $\dfrac{x+5}{2x^2 - 2}$

D. $\dfrac{-x - 5}{2x^2 - 2}$

26) In the following figure, AB is the diameter of the circle. What is the circumference of the circle?

A. 5π

B. 8π

C. 10π

D. 12π

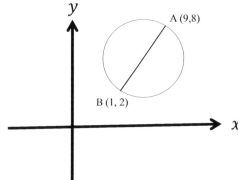

27) 102.5÷0.55 =?

    A. 18.652

    B. 186.36

    C. 186.85

    D. 188.36

28) A circle has a diameter of 14 inches. What is its approximate area?

    A. 120.56

    B. 125.5

    C. 153.86

    D. 200.96

29) If 15 garbage trucks can collect the trash of 45 homes in a day. How many trucks are needed to collect in 120 houses?

    A. 12

    B. 15

    C. 30

    D. 40

30) Solve for $4x^2 - 34 = 66$

    A. ± 3

    B. ± 5

    C. ± 9

    D. ± 10

Use the following table to answer question below.

DANIEL'S BIRD-WATCHING PROJECT	
DAY	NUMBER OF RAPTORS SEEN
Monday	?
Tuesday	7
Wednesday	15
Thursday	23
Friday	7
MEAN	12

31) This table shows the data Daniel collects while watching birds for one week. How many raptors did Daniel see on Monday?

    A. 5

    B. 8

    C. 12

    D. 13

32) A floppy disk shows 856,036 bytes free and 739,352 bytes used. If you delete a file of size 652,159 bytes and create a new file of size 599,986 bytes, how many free bytes will the floppy disk have?

   A. 114,857

   B. 123.986

   C. 221,867

   D. 302,209

33) 6 days 14 hours 42 minutes – 4 days 12 hours 35 minutes =?

   A. 2 days 1 hours 10 minutes

   B. 2 days 2 hours 7 minutes

   C. 2 days 2 hours 13 minutes

   D. 2 days 7 hours 23 minutes

34) If $15 + x^{\frac{1}{2}} = 24$, then what is the value of $8 \times x$?

   A. 321

   B. 456

   C. 550

   D. 648

35) Increased by 35%, the numbers 60 becomes:

   A. 42

   B. 56

   C. 75

   D. 81

36) Which equation represents the statement twice the difference between 8 times h and 4 gives 42.

   A. $\dfrac{8H + 4}{2} = 42$

   B. $2(8H + 4) = 42$

   C. $2(8H - 4) = 42$

   D. $4\dfrac{8H}{2} = 42$

37) A circle is inscribed in a square, as shown below.

The area of the circle is 64π cm². What is the area of the square?

A. 64 cm²

B. 121 cm²

C. 144 cm²

D. 256 cm²

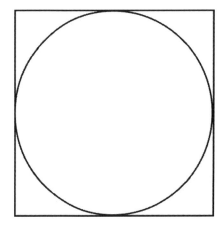

38) The base of a right triangle is 4 foot, and the interior angles are 45-45-90. What is its area?

A. 6 square feet

B. 7 square feet

C. 8 square feet

D. 10 square feet

39) Triangle ABC is graphed on a coordinate grid with vertices at A (-2, -2), B (-3, 8) and C (5, 4). Triangle ABC is reflected over y axes to create triangle A'B'C'.

Which order pair represents the coordinate of C'?

A. (-5, -4)

B. (5, 4)

C. (5, -4)

D. (-5, 4)

40) Which set of ordered pairs represents y as a function of x?

A. {(2, 8), (3, 7), (9, -8), (4, -7)}

B. {(4, 2), (3, -9), (5, 8), (4, 7)}

C. {(9, 12), (5, 7), (6, 11), (5, 18)}

D. {(6, 1), (6, 3), (0, 5), (4, 5)}

41) The width of a box is one fourth of its length. The height of the box is one third of its width. If the length of the box is 24 cm, what is the volume of the box?

   A. 81 cm³
   B. 162 cm³
   C. 252 cm³
   D. 288 cm³

42) A square measures 8 inches on one side. By how much will the area be decreased if its length is increased by 4 inches and its width increased by 2 inches.

   A. 5 sq increased
   B. 6 sq increased
   C. 8 sq increased
   D. 9 sq increased

43) If a box contains red and blue balls in ratio of 3: 5 red to blue, how many red balls are there if 120 blue balls are in the box?

   A. 48
   B. 72
   C. 84
   D. 85

44) How many 5 × 5 squares can fit inside a rectangle with a height of 42 and width of 20?

   A. 33
   B. 35
   C. 37
   D. 42

45) David makes a weekly salary of $250 plus 12% commission on his sales. What will his income be for a week in which he makes sales totaling $1,800?

   A. $328
   B. $452
   C. $466
   D. $485

46) $5x^3y^4 + 14x^2y - (3x^3y^4 - 3x^2y) =$ ___

   A. $8x^3y^4 + 17x^2y$
   B. $2x^3y^4 + 11x^2y$
   C. $2x^3y^4 - 17x^2y$
   D. $2x^3y^4 + 17x^2y$

47) The radius of circle A is fifth times the radius of circle B. If the circumference of circle A is 20π, what is the area of circle B?

   A. 3π
   B. 4π
   C. 9π
   D. 12π

ISEE UPPER LEVEL

Mathematics Practice Test 2

- 47 questions
- **Total time for this section:** 40 Minutes
- **Calculators are not allowed at the test.**

1) $\frac{1}{8b^2} + \frac{1}{4b} = \frac{1}{4b^2}$, then $b = ?$

   A. $-\frac{16}{15}$

   B. $\frac{1}{2}$

   C. $-\frac{15}{16}$

   D. 2

2) $\frac{|10+x|}{8} \leq 5$, then $x = ?$

   A. $-50 \leq x \leq 30$

   B. $-40 \leq x \leq 30$

   C. $-50 \leq x \leq 50$

   D. $-32 \leq x \leq 32$

3) Which of the following points lies on the line $-5x + 3y = 30$?

   A. $(\frac{3}{5}, 11)$

   B. $(-1, 3)$

   C. $(-\frac{3}{5}, 10)$

   D. $(2, 2)$

4) Ella (E) is 5 years older than her friend Ava (A) who is 2 years older than her sister Sofia (S). If E, A and S denote their ages, which one of the following represents the given information?

   A. $\begin{cases} E = A + 5 \\ S = A - 2 \end{cases}$

   B. $\begin{cases} E = A + 5 \\ A = S + 2 \end{cases}$

   C. $\begin{cases} A = E + 5 \\ S = A - 2 \end{cases}$

   D. $\begin{cases} E = A + 5 \\ A = S - 2 \end{cases}$

5) 8 less than twice a positive integer is 84. What is the integer?

   A. 39

   B. 41

   C. 42

   D. 46

6) The cost, in thousands of dollars, of producing x thousands of textbooks is $C(x) = 2x^2 - 3x$. The revenue, also in thousands of dollars, is $R(x) = 36x$. find the profit or loss if 25 textbooks are produced. (profit = revenue − cost)

   A. $2,160 profit

   B. $275 profit

   C. $2,160 loss

   D. $275 loss

7) An angle is equal to one third of its supplement. What is the measure of that angle?

   A. 20

   B. 30

   C. 45

   D. 60

8) Write 46.5 in expanded form, using exponents.

   A. $(4 \times 10) + (5 \times 10^{-1}) + 6$

   B. $(4 \times 10^2) + (5 \times 10^1) - 6$

   C. $(4 \times 10) + (5 \times 10^1) + 6$

   D. $(4 \times 10^1) + (5 \times 10^{-1}) - 6$

9) Right triangle ABC has two legs of lengths 21 cm (AB) and 28 cm (AC). What is the length of the third side (BC)?

   A. 20 cm

   B. 25 cm

   C. 30 cm

   D. 35 cm

10) Simplify $24xy^3 \left(\frac{1}{2}x^2y^2\right)^3 =$

   A. $4x^4y^6$

   B. $4x^8y^6$

   C. $3x^8y^9$

   D. $3x^5y^7$

11) Which is the longest time?

   A. 21 hours

   B. 1530 minutes

   C. 1.5 days

   D. 72000 seconds

12) 3.6 is what percent of 18?

   A. 3.6

   B. 12

   C. 20

   D. 24

13) A company pays its writer $5 for every 600 words written. How much will a writer earn for an article with 1800 words?

   A. $11

   B. $12

   C. $15

   D. $18

14) What is the area of an isosceles right triangle that has one leg that measures 7 cm?

   A. 24.5 cm

   B. 36.5 cm

   C. $6\sqrt{2}$ cm

   D. 72 cm

15) A circle has a diameter of 12 inches. What is its approximate circumference?

   A. 6.28

   B. 25.12

   C. 36.25

   D. 37.7

16) A circular logo is enlarged to fit the lid of a jar. The new diameter is 25% larger than the original. By what percentage has the area of the logo increased?

   A. 54%

   B. 56.25%

   C. 56.35%

   D. 75%

17) What's the area of the non-shaded part of the following figure?

   A. 192

   B. 152

   C. 108

   D. 96

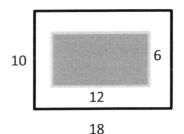

18) 85.5÷6.25?

   A. 13.25

   B. 13.68

   C. 15.68

   D. 16.55

19) A bread recipe calls for $4\frac{1}{4}$ cups of flour. If you only have $2\frac{3}{8}$ cups, how much more flour is needed?

   A. $1\frac{2}{8}$

   B. $\frac{1}{2}$

   C. $1\frac{7}{8}$

   D. $\frac{7}{8}$

20) The equation of a line is given as: y = 8 x − 4. Which of the following points does not lie on the line?

   A. (1, 4)

   B. (−3, −28)

   C. (2, 12)

   D. (2, 7)

21) The drivers at G & G trucking must report the mileage on their trucks each week. The mileage reading of Ed's vehicle was 36,707 at the beginning of one week, and 37,023 at the end of the same week. What was the total number of miles driven by Ed that week?

   A. 46 MILES

   B. 145 MILES

   C. 316 MILES

   D. 1,046 MILES

22) Which equation represents the statement four times the sum of the squares of w and x is 26.

A. 42

B. $4(w^2 + x) = 26$

C. 126

D. $2(w^2 - x) = 26$

23) What is the solution of the following system of equations?
$$\begin{cases} 2x - 3y = 12 \\ x - 2y = 4 \end{cases}$$

A. (-1, 2)

B. (12, 4)

C. (1, 4)

D. (4, -12)

24) What is the area of an isosceles right triangle that has one leg that measures 11 cm?

A. 18 cm

B. 36.5 cm

C. 60.5 cm

D. 72.5 cm

25) Which of the following is a factor of both $2x^2 - 4x$ and $x^2 - 5x + 6$?

A. $(x - 4)$

B. $(x + 4)$

C. $(x - 2)$

D. $(x + 2)$

26) $\frac{12}{32}$ is equal to:

A. 3.6

B. 0.365

C. 0.0375

D. 0.375

27) What's the reciprocal of $\frac{x^4}{32}$?

A. $\frac{32}{x^4} - 1$

B. $\frac{32}{x^{-3}}$

C. $\frac{32}{x^3} + 1$

D. $\frac{32}{x^4}$

28) $\begin{array}{r} 25 \text{ hr. } 36 \text{ min.} \\ \underline{19 \text{ hr. } 15 \text{ min.}} \\ \end{array}$

A. 6 hr. 21 min.

B. 6 hr. 32 min.

C. 6 hr. 15 min.

D. 6 hr. 12 min.

29) A car uses 12 gallons of gas to travel 420 miles. How many miles per gallon does the car get?

    A. 28 miles per gallon

    B. 32 miles per gallon

    C. 35 miles per gallon

    D. 37 miles per gallon

30) Find the perimeter of a rectangle with the dimensions 45 + 39.

    A. 84

    B. 129

    C. 168

    D. 1755

31) If $x + y = 24$, what is the value of $5x + 5y$?

    A. 72

    B. 96

    C. 110

    D. 120

32) Karen is 5 years older than her sister Michelle, and Michelle is 6 years younger than her brother David. If the sum of their ages is 83, how old is Michelle?

    A. 21

    B. 24

    C. 29

    D. 30

33) Ellis just got hired for on-the-road sales and will travel about 2,800 miles a week during an 85-hour work week. If the time spent traveling is $\frac{4}{5}$ of his week, how many hours a week will he be on the road?

    A. Ellis spends about 62 hours of his 85-hour work week on the road.

    B. Ellis spends about 68 hours of his 85-hour work week on the road.

    C. Ellis spends about 72 hours of his 85-hour work week on the road.

    D. Ellis spends about 78 hours of his 85-hour work week on the road.

34) Given that $x = 0.4$ and $y = 8$, what is the value of $5x^2(y - 5)$?

    A. 2.4

    B. 4.2

    C. 11.2

    D. 13.2

35) Mario loaned Jett $1,850 at a yearly interest rate of 8%. After one year what is the interest owned on this loan?

    A. $30

    B. $60

    C. $148

    D. $1,260

36) Calculate the area of a parallelogram with a base of 5 feet and height of 3.6 feet.

   A. 10.8 square feet

   B. 12.2 square feet

   C. 14.8 square feet

   D. 18.0 square feet

37) A tree 48 feet tall casts a shadow 12 feet long. Jack is 8 feet tall. How long is Jack's shadow?

   A. 2 ft

   B. 4 ft

   C. 4.25 ft

   D. 8 ft

38) In a school, the ratio of number of boys to girls is 4:7. If the number of boys is 120, what is the total number of students in the school?

   A. 330

   B. 500

   C. 540

   D. 600

39) A shirt costing $140 is discounted 12%. After a month, the shirt is discounted another 12%. Which of the following expressions can be used to find the selling price of the shirt?

   A. (170) (0.88)

   B. (140) – 140 (0.24)

   C. (140) (0.12) – (140) (0.12)

   D. (140) (0.88) (0.88)

40) What is the area of the shaded region if the diameter of the bigger circle is 15 inches and the diameter of the smaller circle is 7 inches.

   A. 160 π inch²

   B. 176 π inch²

   C. 186 π inch²

   D. 196 π inch²

41) What is the result of the expression?

$$\begin{vmatrix} 5 & -2 \\ 7 & -4 \\ -2 & -1 \end{vmatrix} + \begin{vmatrix} 2 & -1 \\ -5 & 4 \\ 3 & 6 \end{vmatrix} ?$$

A. $\begin{Vmatrix} 7 & -3 \\ 2 & 8 \\ 1 & 5 \end{Vmatrix}$

B. $\begin{Vmatrix} 7 & -3 \\ 2 & 8 \\ 5 & 5 \end{Vmatrix}$

C. $\begin{vmatrix} 7 & -3 \\ 2 & 0 \\ 1 & 5 \end{vmatrix}$

D. $\begin{vmatrix} 0 & -3 \\ -6 & 0 \\ -10 & -3 \end{vmatrix}$

42) There are five equal tanks of water. If $\frac{4}{5}$ of a tank contains 130 liters of water, what is the capacity of the five tanks of water together?

A. 540

B. 655.5

C. 812.5

D. 1,500

43) $(5x - 4)(4x + 2) =$

A. $4x + 8$

B. $20x^2 - 6x - 8$

C. $20x^2 + 18x + 15$

D. $3x^2 + 3$

44) If $x \blacksquare y = \sqrt{x^2 + y}$, what is the value of $6 \blacksquare 13$?

A. $\sqrt{126}$

B. 3

C. 4

D. 7

45) How many square feet of tile is needed for a 26-foot x 26-foot room?

A. 72 square feet

B. 256 square feet

C. 546 square feet

D. 676 square feet

46) The average weight of 22 girls in a class is 55 kg and the average weight of 24 boys in the same class is 70 kg. What is the average weight of all the 46 students in that class?

A. 60

B. 62.82

C. 62.88

D. 64.56

47) If x is 65% percent of 960, what is x?

   A. 185

   B. 369

   C. 624

   D. 720

ISEE Upper Level Practice Tests

Answers and Explanations

ISEE Upper Level Practice Test 2 Answers

1)	C	16)	B	31)	B	46)	D
2)	D	17)	D	32)	A	47)	B
3)	B	18)	A	33)	B		
4)	D	19)	A	34)	D		
5)	B	20)	C	35)	D		
6)	A	21)	D	36)	C		
7)	D	22)	B	37)	D		
8)	B	23)	D	38)	C		
9)	B	24)	D	39)	D		
10)	B	25)	D	40)	A		
11)	C	26)	C	41)	D		
12)	A	27)	B	42)	C		
13)	B	28)	C	43)	B		
14)	D	29)	D	44)	A		
15)	C	30)	B	45)	C		

ISEE Upper Level Practice Test 2 Answers

1)	B	16)	B	31)	D	46)	B
2)	A	17)	C	32)	B	47)	C
3)	A	18)	B	33)	B		
4)	B	19)	C	34)	A		
5)	D	20)	D	35)	C		
6)	B	21)	C	36)	D		
7)	C	22)	B	37)	A		
8)	A	23)	B	38)	A		
9)	D	24)	C	39)	D		
10)	C	25)	C	40)	B		
11)	C	26)	D	41)	C		
12)	C	27)	D	42)	C		
13)	C	28)	A	43)	B		
14)	A	29)	C	44)	D		
15)	D	30)	C	45)	D		

ISEE Upper Level Practice Tests 1

Answers and Explanations

1) **Choice C is correct**

   $|15 - (18 \div |2 - 8|)| = |15 - (18 \div 6)| = |15 - 3| = 12$

2) **Choice D is correct**

   Use FOIL (First, Out, In, Last) method.

   $(x - 4)(2x + 2) = 2x^2 + 2x - 8x - 8 = 2x^2 - 6x - 8$

3) **Choice B is correct**

   Solve for x. $-4 \leq 4x - 5 < 7 \Rightarrow$ (add 5 all sides) $5 - 4 \leq 4x - 5 + 5 < 7 + 5 \Rightarrow$

   $1 \leq 4x < 12 \Rightarrow$ (divide all sides by 4) $\frac{1}{4} \leq x < 3$, x is between $\frac{1}{4}$ and 3. Choice B represent this inequality.

4) **Choice D is correct**

   $x_{1,2} = \frac{-b \pm \sqrt{b^2 - 4ac}}{2a}$, $ax^2 + bx + c = 0$, $3x^2 + 8x + 4 = 0 \Rightarrow$ then: a = 3, b = 8 and c = 4

   $x = \frac{-8 + \sqrt{8^2 - 4.3.4}}{2.3} = -\frac{2}{3}$, $x = \frac{-8 - \sqrt{8^2 - 4.3.4}}{2.3} = -2$

5) **Choice B is correct.**

   $\frac{8}{23}$ = 0.347 and 45% = 0.45 therefore x should be between 0.0.347 and 0.45

   Choice B. $\frac{8}{19}$ = 0.421 is between 0.347 and 0.45

6) **Choice A is correct.**

   A linear equation is a relationship between two variables, x and y, and can be written in the form of $y = mx + b$.

   A non-proportional linear relationship takes on the form $y = mx + b$, where b ≠ 0 and its graph is a line that does not cross through the origin.

7) **Choice D is correct.**

   Translated 6 units down and 5 units to the left means: $(x. y) \Rightarrow (x - 5, y - 6)$

8) **Choice B is correct.**

Write the proportion and solve for missing side.

$$\frac{\text{Smaller triangle height}}{\text{Smaller triangle base}} = \frac{\text{Bigger triangle height}}{\text{Bigger triangle base}} \Rightarrow \frac{100cm}{150cm} = \frac{100+280cm}{x} \Rightarrow x = 570 \text{ cm}$$

9) Choice B is correct.

$0.00000625 = \dfrac{6.25}{1000000} \Rightarrow 6.25 \times 10^{-6}$

10) Choice B is correct

The ratio of boy to girls is 7:5. Therefore, there are 7 boys out of 12 students. To find the answer, first divide the total number of students by 12, then multiply the result by 7.

$540 \div 12 = 45 \Rightarrow 120 \times 7 = 315$

11) Choice C is correct

Probability = (number of desired outcomes)/ (number of total outcomes)

In this case, a desired outcome is selecting either a red or a yellow marble. Combine the number of red and yellow marbles: 7 + 6 = 13 and divide this by the total number of marbles: 5 + 7 + 6 = 18. The probability is $\dfrac{13}{18}$.

12) Choice A is correct

Emily = Daniel, Emily = 9 Claire, Daniel = 24 + Claire, Emily = Daniel → Emily = 20 + Claire

Emily = 9 Claire → 9 Claire = 24 + Claire → 9 Claire − Claire = 24, 8 Claire = 24, Claire = 3

13) Choice B is correct

$\dfrac{1}{4}$ of the distance is $4\dfrac{2}{5}$ miles. Then: $\dfrac{1}{4} \times 4\dfrac{2}{5} = \dfrac{1}{4} \times \dfrac{22}{5} = \dfrac{22}{20}$

Converting $\dfrac{22}{20}$ to a mixed number gives: $\dfrac{22}{20} = 1\dfrac{2}{20} = 1\dfrac{1}{10}$

14) Choice D is correct

Use Pythagorean theorem: $a^2 + b^2 = c^2 \to s^2 + h^2 = (10s)^2 \to (6s)^2 + h^2 = 100s^2$

Subtracting s^2 from both sides gives: $h^2 = 64s^2$,

Square roots of both sides: $h = \sqrt{64s^2} = 8s$

15) Choice C is correct

$5 − 12 \div (2^4 \div 4) = 5 − 12 \div (16 \div 4) = 5 − 12 \div (4) = 5-3 =2$

16) Choice C is correct

The area of the trapezoid is: $Area = \frac{1}{2}h(b_1 + b_2) = \frac{1}{2}(x)(14 + 10) = 36$

$\rightarrow 12x = 36 \rightarrow x = 3$, $y = \sqrt{4^2 + 3^2} = \sqrt{16 + 9} = \sqrt{25} = 5$

The perimeter of the trapezoid is: $10 + 4 + +10 + 3 + 5 = 32$

17) Choice D is correct

$$average = \frac{sum}{total} = \frac{40 + 30 + 38}{3} = \frac{108}{3} = 36$$

18) Choice A is correct

Area = w × h, Area = 212 × 58 = 12,296

19) Choice A is correct

$(6.2+6.3+6.5)\, x = x+2$, $19x = x + 2$, Then $x = \frac{1}{9}$

20) Choice C is correct

16% of x = 15.4, $x = \frac{16}{100}x = 15.4$, $x = \frac{16 \times 15.4}{100} = 96.25$

21) Choice D is correct

$\frac{9}{30} = 0.3$

22) Choice B is correct

$\frac{240}{8} < x < \frac{320}{8}$, $30 < x < 40$, Then: Only choice b is correct

23) Choice D is correct

$75 \div \frac{1}{6} = \frac{\frac{75}{1}}{\frac{1}{6}} = 75 \times 6 = 450$

24) Choice D is correct

For sum of 5: (1 & 4) and (4 & 1), (2 & 3) and (3 & 2), therefore we have 4 options.

For sum of 7: (1 & 6) and (6 & 1), (2 & 5) and (5 & 2), (3 & 4) and (4 & 3) we have 6 options.

To get a sum of 6 or 9 for two dice: 6 + 4 = 10

Since, we have 6 × 6 = 36 total number of options, the probability of getting a sum of 4 and 6 is 10 out of 36 or $\frac{10}{36} = \frac{5}{18}$.

25) Choice D is correct

Simplify: $\dfrac{\frac{1}{4}-\frac{x+7}{8}}{\frac{x^2}{4}-\frac{1}{4}} = \dfrac{\frac{1}{4}(1-\frac{x+7}{2})}{\frac{x^2-1}{4}} = \dfrac{(1-\frac{x+7}{2})}{x^2-1}$, ⇒Simplify: $1-\dfrac{x+7}{2} = \dfrac{2-x-7}{2} = \dfrac{-x-5}{2}$,

Then: $\dfrac{(1-\frac{x+7}{2})}{x^2-1} = \dfrac{\frac{-x-5}{2}}{x^2-1} = \dfrac{-x-5}{2(x^2-1)} = \dfrac{-x-5}{2x^2-2}$

26) Choice C is correct

The distance of A to B on the coordinate plane is: $\sqrt{(x_1-x_2)^2+(y_1-y_2)^2} = \sqrt{(9-1)^2+(8-2)^2} = \sqrt{8^2+6^2}, = \sqrt{64+36} = \sqrt{100} = 10$

The diameter of the circle is 10 and the radius of the circle is 5. Then: the circumference of the circle is: $2\pi r = 2\pi(5) = 10\pi$

27) Choice B is correct

$102.5 \div 0.55 = 186.36$

28) Choice C is correct

Diameter = 14, then: Radius = 7, Area of a circle = πr^2 ⇒ A = $3.14(7)^2 = 153.86$

29) Choice D is correct

$\dfrac{15}{45} = \dfrac{x}{120} \to x = \dfrac{15 \times 120}{45} = 40$

30) Choice B is correct

$4x^2 - 34 = 66$, $4x^2 = 100$, $x^2 = 25$, $x = \pm 5$

31) Choice B is correct

$\dfrac{x+7+15+23+7}{5} = 12 \to x + 52 = 60 \to x = 62 - 52 = 8$

32) Choice A is correct

The difference of the file added, and the file deleted is: $793,352 - 652,159 + 599,986 = 741,179$, $856,036 - 741,179 = 114,857$

33) Choice B is correct

6 days 14 hours 42 minutes − 4 days 12 hours 35 minutes = 2 days 2 hours 7 minutes

34) Choice D is correct

$x^{\frac{1}{2}}$ equals to the root of x. Then: $15 + x^{\frac{1}{2}} = 24 \to 15 + \sqrt{x} = 24 \to \sqrt{x} = 9 \to x = 81$

$x = 81$ and $8 \times x$ equals: $8 \times 81 = 648$

35) Choice D is correct

0.35% of 60 = 21, 60 + 21= 81

36) Choice C is correct

2(8h - 4) = 42

37) Choice D is correct

The area of the circle is 64π cm², then, its diameter is 8cm.

$$area\ of\ a\ circle = \pi r^2 = 64\pi \rightarrow r^2 = 64 \rightarrow r = 8$$

Radius of the circle is 8 and diameter is twice of it, 16.

One side of the square equals to the diameter of the circle. Then:

$$Area\ of\ square = side \times side = 16 \times 16 = 256$$

38) Choice C is correct

Formula of triangle area = ½ (base × height)

Since the angles are 45-45-90, then this is an isosceles triangle, meaning that the base and height of the triangle are equal. Triangle area = ½ (base × height) = ½ (4 × 4) = 8

39) Choice D is correct.

When a point is reflected over y axes, the (x) coordinate of that point changes to $(-x)$ while its x coordinate remains the same. C (5, 4) → C' (−5, 4)

40) Choice A is correct.

A set of ordered pairs represents y as a function of x if: $x_1 = x_2 \rightarrow y_1 = y_2$

In choice B: (4, 2) and (4, 7) are ordered pairs with same x and different y, therefore y isn't a function of x.

In choice C: (5, 7) and (5, 18) are ordered pairs with same x and different y, therefore y isn't a function of x.

In choice D: (6, 1) and (6, 3) are ordered pairs with same x and different y, therefore y isn't a function of x.

41) Choice D is correct

If the length of the box is 24, then the width of the box is one fourth of it, 6, and the height of the box is 2 (one third of the width). The volume of the box is: $V = lwh$ = (24)(6) (2) = 288

42) Choice C is correct

The area of the square is 64 square inches. $Area\ of\ square = side \times side = 8 \times 8 = 64$

The length of the square is increased by 4 inches and its width decreased by 2 inches. Then, its area equals: $Area\ of\ rectangle = width \times ength = 12 \times 6 = 72$

The area of the square will be increased by 3 square inches. $72 - 64 = 8$

43) Choice B is correct

Write a proportion and solve. $\frac{3}{5} = \frac{x}{120}$

Use cross multiplication: $5x = 360 \rightarrow x = 72$

44) Choice A is correct

Number of squares equal to: $\frac{42 \times 20}{5 \times 5} = 33.6$

45) Choice C is correct

David's weekly salary is $250 plus 12% of $1,800. Then: $12\%\ of\ 1,800 = 0.12 \times 1,800 = 216$, $250 + 216 = 466$

46) Choice D is correct

$$5x^3y^4 + 14x^2y - (3x^3y^4 - 3x^2y) = 2x^3y^4 + 17x^2y$$

47) Choice B is correct

Let P be circumference of circle A, then; $2\pi r_A = 20\pi \rightarrow r_A = 10$

$r_A = 5r_B \rightarrow r_B = \frac{10}{5} = 2 \rightarrow$ Area of circle B is; $\pi r_B^2 = 4\pi$

ISEE Upper Level Practice Tests 2

Answers and Explanations

1) Choice B is correct

Subtract $\frac{1}{8b}$ and $\frac{1}{8^2}$ from both sides of the equation. Then:

$$\frac{1}{8b^2} + \frac{1}{4b} = \frac{1}{4b^2} \rightarrow \frac{1}{8b^2} - \frac{1}{4b^2} = -\frac{1}{4b}$$

Multiply both numerator and denominator of the fraction $\frac{1}{4b^2}$ by 2. Then: $\frac{1}{8b^2} - \frac{2}{8b^2} = -\frac{1}{4b}$, Simplify the first side of the equation: $-\frac{1}{8b^2} = -\frac{1}{4b}$

Use cross multiplication method: $4b = 8b^2 \rightarrow 1 = 2b \rightarrow b = \frac{1}{2}$

2) Choice A is correct

First, multiply both sides of inequality by 8. Then: $\frac{|10+x|}{8} \leq 5 \to |10+x| \leq 40$

Since $10 + x$ can be positive or negative, then: $10 + x \leq 40$ or $10 + x \geq -40$

Then: $x \leq 30$ or $x \geq -50$, Choice B is correct.

3) **Choice A is correct.**

   Plug in each pair of numbers in the equation. The answer should be 14.

   A. $(\frac{3}{5}, 11)$: $-5(\frac{3}{5}) + 3(11) = 30$ Yes!
   B. $(-1, 3)$: $-5(-1) + 3(3) = 14$ No!
   C. $(-\frac{3}{5}, 12)$: $-5(-\frac{3}{5}) + 3(12) = 39$ No!
   D. $(2, 2)$: $-5(2) + 3(2) = -4$ No!

4) **Choice B is correct**

   E = 5 + A, A = S + 2

5) **Choice D is correct**

   Let x be the integer. Then: $2x - 8 = 84$
   Add 8 both sides: $2x = 92$, Divide both sides by 2: $x = 46$

6) **Choice B is correct**

   Plug in the value of $x = 25$ into both equations. Then:

   $C(x) = 2x^2 - 3x = 2(25)^2 - 3(25) = 1250 - 75 = 1175$, $R(x) = 36x = 36 \times 25 = 900$, $1,175 - 900 = 275$

7) **Choice C is correct**

   The sum of supplement angles is 180. Let x be that angle. Therefore, $x + 3x = 180$

   $4x = 180$, divide both sides by 4: $x = 45$

8) **Choice A is correct**

   Let's review the choices provided: A. $(4 \times 10) + (5 \times 10^{-1}) + 6 = 40 + 0.5 + 6 = 46.5$

   B. $(4 \times 10^2) + (5 \times 10^1) - 6 = 400 + 50 - 6 = 444$

   C. $(4 \times 10) + (5 \times 10^1) + 6 = 400 + 50 + 6 = 456$

   D. $(4 \times 10^1) + (5 \times 10^{-1}) - 6 = 40 + 0.5 - 6 = 34.5$, Only choice C equals to 46.5.

9) **Choice D is correct**

   Use Pythagorean Theorem: $a^2 + b^2 = c^2$, $21^2 + 28^2 = c^2 \Rightarrow 1225 = c^2 \Rightarrow c = 35$

10) Choice C is correct

Simplify. $24xy^3 \left(\frac{1}{2}x^2y^2\right)^3 == 24x^2y^3 \left(\frac{1}{8}x^6y^6\right) = 3x^8y^9$

11) Choice C is correct

21 hours = 75,600 seconds, 1530 minutes = 91,800 seconds, 1.5 days = 36 hours = 129,600 seconds, 72000 seconds

12) Choice C is correct

$x\%$ 18 = 3.6, $\frac{x}{100}$ 18 = 3.6 → $x = \frac{3.6 \times 100}{18} = 20$

13) Choice C is correct

$\frac{5}{600} = \frac{x}{1800}$, $x = \frac{5 \times 1800}{600} = 15$

14) Choice A is correct

First draw an isosceles triangle. Remember that two sides of the triangle are equal.

Isosceles right triangle

Let put a for the legs. Then:

$a = 7 \Rightarrow$ area of the triangle is $= \frac{1}{2}(7 \times 7) = \frac{49}{2} = 24.5\ cm^2$

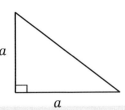

15) Choice D is correct

C = 2πr, C = 2π × 6 = 12π, π = 3.14   →   C = 12π = 37.7

16) Choice B is correct

Area of a circle equals: $A = \pi r^2$

The new diameter is 25% larger than the original then the new radius is also 25% larger than the original. 25% larger than r is $1.25r$. Then, the area of larger circle is:

$A = \pi r^2 = \pi(1.25r)^2 = \pi(1.5625r^2) = 1.5625\pi r^2$, $1.56.25\pi r^2$ is 56.25% bigger than πr^2.

17) Choice C is correct

The area of the non-shaded region is equal to the area of the bigger rectangle subtracted by the area of smaller rectangle. Area of the bigger rectangle = 10 × 18 = 180

Area of the smaller rectangle = 12 × 6 = 72, Area of the non-shaded region = 180 − 72 = 108

18) Choice B is correct

85.5 ÷ 6.25 = 13.68

19) Choice C is correct

$$4\frac{1}{4} - 2\frac{3}{8} = \frac{17}{4} - \frac{19}{8} = \frac{34-19}{8} = \frac{15}{8} = 1\frac{7}{8}$$

20) Choice D is correct

Let's review the choices provided. Put the values of x and y in the equation.

A. (1, 4)   ⇒ $x = 1$ ⇒ $y = 4$ This is true!

B. (−3, −28)   ⇒ $x = -3$ ⇒ $y = -24$ This is true!

C. (2, 12)   ⇒ $x = 2$ ⇒ $y = 12$ This is true!

D. (2, 7)   ⇒ $x = 2$ ⇒ $y = 12$ This is not true!

21) Choice C is correct

To find total number of miles driven by Ed that week, you only need to subtract 37,707 from 36,023. 37,023 − 36,707 = 316

22) Choice B is correct

$4(w^2 + x) = 26$

23) Choice B is correct

$\begin{cases} 2x - 3y = 12 \\ x - 2y = 4 \end{cases}$ ⇒ Multiplication (−2) in first equation ⇒ $\begin{cases} 2x - 3y = 12 \\ -2x + 4y = -8 \end{cases}$

Add two equations together ⇒ y = 4 ⇒ then: x = 12

24) Choice C is correct

First draw an isosceles triangle. Remember that two sides of the triangle are equal.

Isosceles right triangle

Let put a for the legs. Then:

$a = 11$ ⇒ area of the triangle is $= \frac{1}{2}(11 \times 11) = \frac{121}{2} = 60.5 \ cm^2$

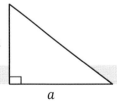

25) Choice C is correct

Factor each trinomial $2x^2 - 4x$ and $x^2 - 5x + 6$, $2x^2 - 4x$ ⇒ $2x(x - 2)$

$x^2 - 5x + 6$ ⇒ $(x - 2)(x - 3)$

26) Choice D is correct

$\frac{12}{32} = 0.375$

27) Choice D is correct

$\frac{x^4}{32} \Rightarrow$ reciprocal is: $\frac{32}{x^4}$

28) Choice A is correct

  25 hr. 36 min.
  19 hr. 15 min.
   6hr. 21min

29) Choice C is correct

$\frac{420}{12} = 35$

30) Choice C is correct

$Perimeter\ of\ a\ rectangle = 2(width + length) = 2(45 + 39) = 168$

31) Choice D is correct

$x + y = 24$, Then: $5x + 5y = 24 \times 5 = 120$

32) Choice B is correct

Michelle = Karen − 5, Michelle = David − 6, Karen + Michelle + David = 83

Karen -5 = Michelle ⇒ Karen = Michelle +5, Karen + Michelle + David = 83

Now, replace the ages of Karen and David by Michelle. Then: Michelle + 5 + Michelle + Michelle + 6 = 61, 3Michelle + 11 = 83 ⇒ 3Michelle = 83 − 11, 3Michelle = 72, Michelle = 24

33) Choice B is correct

Ellis travels $\frac{4}{5}$ of 85 hours. $\frac{4}{5} \times 85 = 68$, Ellis will be on the road for 68 hours.

34) Choice A is correct

$5x^2(y − 5) = 5(0.4)^2(8 − 5) = 5\ (0.16)\ (3) = 2.4$

35) Choice C is correct

Use interest rate formula: $Interet = rncipal \times rate \times time = 1,850 \times 0.08 \times 1 = 148$

36) Choice D is correct

A = bh, A = 5 × 3.6 = 18

37) Choice A is correct

Write a proportion and solve for the missing number. $\frac{48}{12} = \frac{8}{x} \to 48x = 8 \times 12 = 96$

$$48x = 96 \to x = \frac{96}{48} = 2$$

38) Choice A is correct.

The ratio of boys to girls is 4:7. Therefore, there are 4 boys out of 11 students. To find the answer, first divide the number of boys by 4, then multiply the result by 11.

120 ÷ 4 = 30 ⇒ 30 × 11 = 330

39) Choice D is correct

To find the discount, multiply the number by (100% − rate of discount).

Therefore, for the first discount we get: (140) (100% − 12%) = (140) (0.88)

For the next 12 % discount: (140) (0.88) (0.88)

40) Choice B is correct.

To find the area of the shaded region subtract smaller circle from bigger circle.

$S_{bigger} - S_{smaller} = \pi (r_{bigger})^2 - \pi (r_{smaller})^2 \Rightarrow S_{bigger} - S_{smaller} = \pi (15)^2 - \pi (7)^2$

⇒ 225 π − 49π = 176 π

41) Choice C is correct.

To add two matrices, first we need to find corresponding members from each matrix.

$$\begin{vmatrix} 5 & -2 \\ 7 & -4 \\ -2 & -1 \end{vmatrix} + \begin{vmatrix} 2 & -1 \\ -5 & 4 \\ 3 & 6 \end{vmatrix} = \begin{vmatrix} 7 & -3 \\ 2 & 0 \\ 1 & 5 \end{vmatrix}$$

ISEE Upper Level Math Preparation Exercise Book

42) Choice C is correct

Let x be the capacity of one tank. Then, $\frac{4}{5}x = 130 \rightarrow x = \frac{130 \times 5}{4} = 162.5$ Liters

The amount of water in five tanks is equal to: $5 \times 162.5 = 812.5$ Liters

43) Choice B is correct

Use FOIL (First, Out, In, Last), $(5x - 4)(4x + 2) = 20x^2 + 10x - 16x - 8 = 20x^2 - 6x - 8$

44) Choice D is correct

Plug in the values of x and y in the equation: $6 \blacksquare 13 = \sqrt{6^2 + 13} = \sqrt{36 + 13} = \sqrt{49} = 7$

45) Choice D is correct

The area of a 26 feet x 26 feet room is 676 square feet.
$26 \times 26 = 676$

46) Choice B is correct

Average = $\frac{\text{sum of terms}}{\text{number of terms}}$, The sum of the weight of all girls is: $22 \times 55 = 1210$ kg

The sum of the weight of all boys is: $24 \times 70 = 1680$ kg

The sum of the weight of all students is: $1680 + 1210 = 2890$ kg, Average = $\frac{2890}{46} = 62.82$

47) Choice C is correct

$\frac{65}{100} \times 960 = x$

www.EffortlessMath.com

... So Much More Online!

✓ FREE Math lessons

✓ More Math learning books!

✓ Mathematics Worksheets

✓ Online Math Tutors

Need a PDF version of this book?

Send email to: info@EffortlessMath.com

Made in the USA
Lexington, KY
21 May 2019